ELECTRICAL, PLUMBING, INSULATION AND THE INTERIOR

THE ILLUSTRATED HOME SERIES

MECHANICS' INSTITUTE

ALSO AVAILABLE IN THE ILLUSTRATED HOME SERIES

STRUCTURE, ROOFING AND THE EXTERIOR

HEATING AND AIR CONDITIONING

ALSO BY ALAN CARSON AND ROBERT DUNLOP

INSPECTING A HOUSE: A GUIDE FOR BUYERS, OWNERS, AND RENOVATORS

ELECTRICAL, PLUMBING, INSULATION AND THE INTERIOR

THE ILLUSTRATED HOME SERIES

Carson Dunlop & Associates

Stoddart

Published in 2000 by Stoddart Publishing Co. Limited
34 Lesmill Road, Toronto, Canada M3B 2T6
180 Varick Street, 9th Floor, New York, New York 10014

Distributed in Canada by:
General Distribution Services Ltd.
325 Humber College Blvd., Toronto, Ontario M9W 7C3
Tel. (416) 213-1919   Fax (416) 213-1917
Email cservice@genpub.com

Distributed in the United States by:
General Distribution Services Inc.
PMB 128, 4500 Witmer Industrial Estates, Niagara Falls, New York 14305-1386
Toll-free Tel.1-800-805-1083   Toll-free Fax 1-800-481-6207
Email gdsinc@genpub.com

04   03   02   01   00   1   2   3   4   5

Canadian Cataloguing in Publication Data

Carson, Alan
Electrical, plumbing, insulation and the interior
(Illustrated home series)
ISBN 0-7737-6147-0
1. Dwellings — Remodeling — Pictorial works. 2. Dwellings — Maintenance and repair — Pictorial works.
3. Dwellings — Electric equipment — Pictorial works. 4. Plumbing — Pictorial works.
5. Dwellings — Insulation — Pictorial works.
I. Dunlop, Robert. II. Title. III. Series: Carson, Alan. Illustrated home series.
TH4816.C37 2000   690'.8'0286   C00-931477-6

U.S. Cataloguing in Publication Data
(Library of Congress Standards)

Carson, Alan.
Electrical, plumbing, insulation and the interior / Alan Carson and Robert Dunlop. — 1st ed.
[128]p: ill. ; cm. (The Illustrated Home Series)
ISBN 0-7737-61470 (pbk.)
1. Dwellings — Maintenance and repair — Amateurs' manuals.
2. Electric wiring — Amateurs' manuals. 3. Plumbing — Repairing — Amateurs' manuals. 4. Dwellings —
Insulation — Amateurs' Manuals . I. Dunlop, Robert. II. Title. III. Series
690 21 2000   CIP

Cover Design: Tannice Goddard
Text Illustrations: VECTROgraphics
Text Design: Neglia Design Inc./Tannice Goddard

THE CANADA COUNCIL | LE CONSEIL DES ARTS
FOR THE ARTS | DU CANADA
SINCE 1957 | DEPUIS 1957

*We acknowledge for their financial support of our
publishing program the Canada Council, the Ontario Arts
Council, and the Government of Canada through the
Book Publishing Industry Development Program (BPIDP).*

Printed and bound in Canada

PART FOUR: THE INTERIOR

INTERIORS

INTRODUCTION

At Carson Dunlop & Associates, a consulting engineering firm, our principle business is home inspection. A few years ago, we set out to build a distance education product for those planning to enter the home inspection business. We developed the Carson Dunlop Home Study System, the most comprehensive home inspection training program in existence.

Early in the development process, we were unable to find existing illustrations that would *show* our students what we were explaining to them in the text. Even when we could find good illustrations, they were limited to a specific topic, and similar illustrations for other subjects couldn't be found at all.

Enter Peter Yeates, an engineer at Carson Dunlop & Associates, and owner of a graphics company called VectroGraphics. He solved our problem by designing over 1,500 illustrations for us. Peter is one of a handful of people on the planet who combines the technical expertise, computer skills, and aesthetics sense needed to create these illustrations.

Not only have these illustrations proven to be an effective tool in training home inspectors, but they have turned out to be very useful during home inspections. They allow us to explain situations to our clients, the potential homeowner.

The comments from students, other home inspectors, and, most importantly, the home buying public, has led us to assemble the illustrations from the Home Study System and publish them as a series of books. Finally, you don't have to feel guilty just looking at the pictures.

THE INTERIOR

ELECTRICITY

Nothing can kill you faster than electricity. The electrical illustrations in this book will clearly show you some of the dos and don'ts. Starting with the basics of electricity, and dealing with the electrical service from the point where it enters the house, the illustrations will take you from the electrical panel through to the methods of distributing power throughout the house.

PLUMBING

The plumbing illustrations are divided into four main categories: supply, water heaters, waste, and fixtures and faucets.

Supply

The supply plumbing brings the water into the house, and distributes fresh, clean water to kitchens, bathrooms, and outdoor faucets.

Water Heaters

In addition to getting the water in and out of the house, water also has to be heated — unless people want to take cold showers! The section of illustrations on water heaters fits the bill.

Waste

Drain, waste, and vent plumbing (DWV) take the dirty water away. The DWV plumbing is much harder to install than the supply plumbing because the water coming into the house is under pressure and the water leaving the house only does so by gravity. If the waste plumbing system is installed incorrectly, slow drains and sewer backup will occur in the house, possibly with sewer odors. The plumbing illustrations will outline how it should be done.

Fixtures and Faucets

Illustrations for dishwashers, faucets, toilets, bidets, bathtubs, shower stalls, etc. are also included for your reference.

INSULATION

Insulation is not a particularly dynamic subject; however, a lack of insulation, or improperly installed insulation or vapor barriers, can result in serious damage to a home. The illustrations in this section will show where insulation should and should not be. The importance of ventilation is also illustrated.

We've included a number of illustrations depicting

heat recovery ventilators (HRVs) and how they work. HRVs allow stale air within the house to be exhausted as fresh air is brought in to replace it. The HRV takes the heat from the exhaust air and uses it to preheat the fresh air. Without these illustrations, the interior workings of an HRV would remain a mystery to most people.

THE THINGS WE SEE

Finally, the interior illustrations show the components of a house that people are most familiar with — floor systems, walls, ceilings, stairs, skylights, solariums, etc.

People are most comfortable with the interior of the house, so this is perhaps the shortest section of the book.

Enjoy!

INSTRUCTIONS FOR USE

You'll notice that there's a table of contents at the beginning of each part of this book. The heading of each illustration is listed next to the number of the illustration. Simply flip through the book until you find the illustration you want to look at.

PART 1

ELECTRICAL

THE DISTRIBUTION SYSTEM

BRANCH CIRCUIT WIRING (DISTRIBUTION WIRING)

KNOB AND TUBE WIRING

ALUMINUM WIRE

LIGHTS, OUTLETS, SWITCHES AND JUNCTION BOXES

Where does the electrical inspection start?

the service entrance conductors are part of the inspection, but the service drop is typically the responsibility of the electrical utility

service drop from utility pole

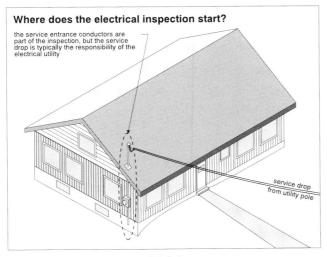

0001

Electricity - basic concepts

electrons (electricity) travelling along a wire

electricity flow can be compared to water flow - if pressure is applied at one end of a pipe (or wire) then, water (or electricity) will flow out the other end

water flowing through a pipe

0002

Electrical potential

the water has potential because (given a chance) it will seek the point of lowest potential i.e., spill onto the floor

a live wire also has potential

it would like to reach ground and will travel through anything with conductivity (e.g., your body) to get there

live wire

ground = zero potential

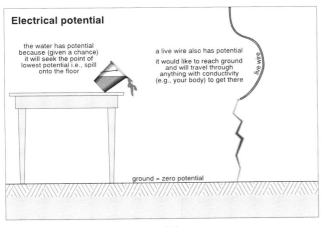

0003

Electrical resistance

resistance of light bulb filament is high so a lot of heat and light are produced as the electricity forces its way through

resistance along circuit wire is quite low

the light bulb is a resistor

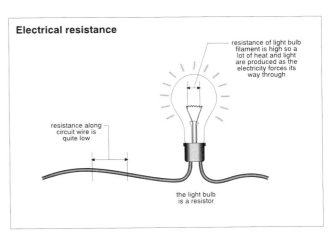

0004

Current flow = amperage

to ground

air gap very high resistance

voltage source

A no current flow no amperage

to ground

large resistor

voltage source

B low current flow low amperage

to ground

small resistor

voltage source

C higher current flow higher amperage

to ground

wires crossed (short circuit) same as no resistor

voltage source

D very high current flow very high amperage

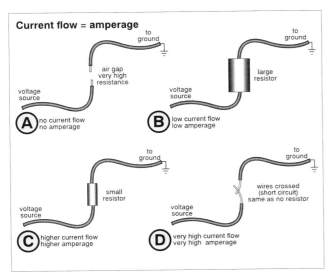

0005

120/240 volts

the neutral wire is typically grounded at the pole and the house and often acts as a support wire for the two "hot" wires

transformer

service drop

utility pole

service panel

to ground

"black" neutral "red"

120 volts   120 volts

to ground

240 volts

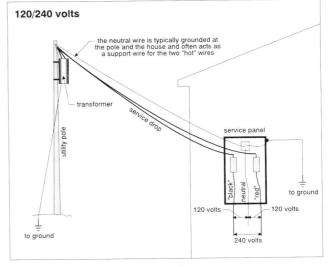

0006

Color coding for typical 120 volt circuit

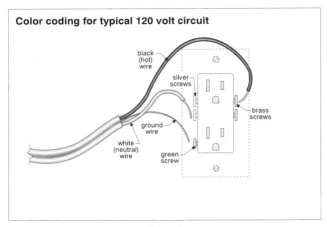

black (hot) wire

silver screws

brass screws

ground wire

white (neutral) wire

green screw

0007

A simple electrical circuit

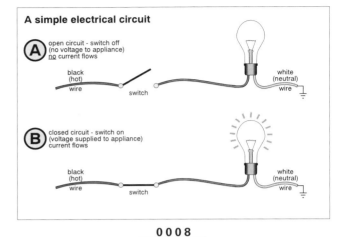

A open circuit - switch off (no voltage to appliance) <u>no</u> current flows

black (hot) wire

switch

white (neutral) wire

B closed circuit - switch on (voltage supplied to appliance) current flows

black (hot) wire

switch

white (neutral) wire

0008

Short circuit

a short circuit occurs when a voltage source has a low resistance path to ground

the result is a very large current flow and the wires can overheat

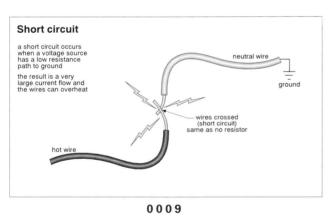

neutral wire

ground

wires crossed (short circuit) same as no resistor

hot wire

0009

Fuses and breakers

fuses are designed to protect the wire from overheating in the case of a short circuit (or overload)

a metal link in the fuse melts (shutting down the circuit) if a current greater than the fuse rating tries to flow through the circuit

neutral wire

wires crossed (short circuit)

hot wire

15 AMP FUSE

fuse/breaker size must match wire size

for example, a 15 amp fuse is used with 14 gauge copper wire

0010

Fuses provide protection

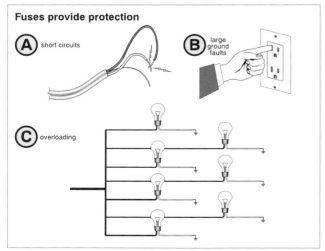

A short circuits

B large ground faults

C overloading

0011

Household wiring is done in parallel

even if several fixtures are controlled by one switch, they are installed in parallel (see upper right)

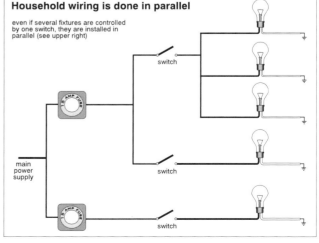

switch

15 AMP FUSE

switch

main power supply

switch

15 AMP FUSE

switch

0012

Damaged wire

wire can easily be nicked when the insulation is being stripped

this creates a localized hotspot that is a fire hazard

wire nicking is more likely when dealing with aluminum wiring since it's softer

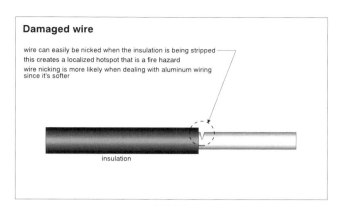

insulation

0 0 1 3

Service drop

the service drop is the wires running from the utility pole to the point of connection to the house

these are sometimes referred to as overhead wires or overhead service

the service drop terminates at the drip loop

an undergound service has buried service laterals running from the utility to the service entrance conductors

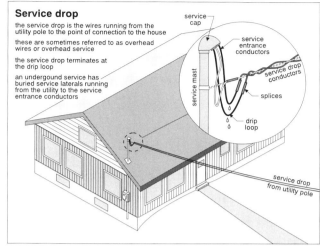

service cap

service entrance conductors

service mast

service drop conductors

splices

drip loop

service drop from utility pole

0 0 1 4

Underground electrical service

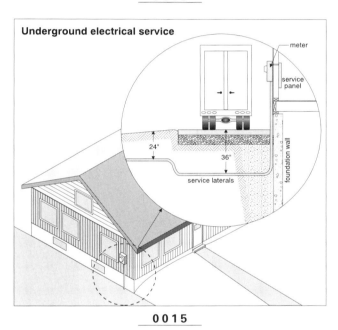

meter

service panel

24"

36"

service laterals

foundation wall

0 0 1 5

Service drop clearances (United States)

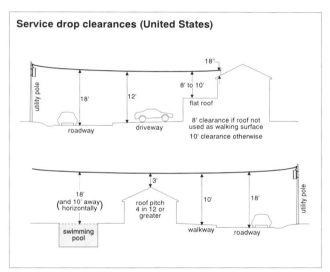

18"

8' to 10'

flat roof

utility pole

18'

12'

roadway

driveway

8' clearance if roof not used as walking surface

10' clearance otherwise

3'

18' (and 10' away horizontally)

roof pitch 4 in 12 or greater

10'

18'

utility pole

swimming pool

walkway

roadway

0 0 1 6

Service drop clearances (Canada)

24"

8' to 10'

flat roof

utility pole

20'

15'

roadway

driveway

15'

20'

utility pole

walkway

roadway

running service wires above roofs is discouraged

0 0 1 7

Wire clearances around windows

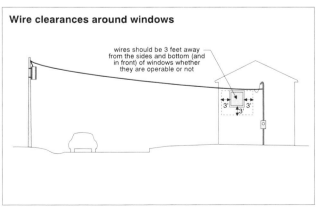

wires should be 3 feet away from the sides and bottom (and in front) of windows whether they are operable or not

3'

3'

0 0 1 8

Wires not well secured to the house

failed service drop connection to service mast results in service drop conductors being pulled tight - the drip loop disappears

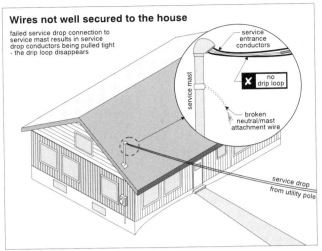

service entrance conductors

service mast

✗ no drip loop

broken neutral/mast attachment wire

service drop from utility pole

0019

Inadequate wire clearance over roofs

(A) UNITED STATES
check to ensure that the clearances indicated below have been provided

(B) CANADA
installing wires over roofs is not generally allowed except by special permission

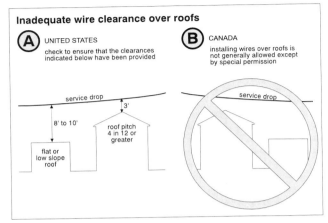

service drop

8' to 10'

flat or low slope roof

3'

roof pitch 4 in 12 or greater

service drop

0020

Drip loop

the primary purpose of the drip loop is to prevent water that runs down the service drop from getting into the service mast (and ultimately into the service panel)

the support wire for the two hot wires can also do double duty as the neutral connection for the house and is grounded back at the power pole

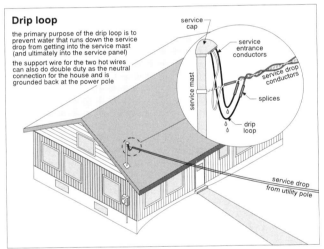

service cap

service entrance conductors

service mast

service drop conductors

splices

drip loop

service drop from utility pole

0021

Service entrance wires

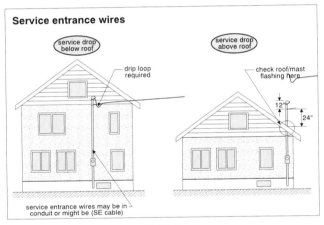

service drop below roof

drip loop required

service entrance wires may be in conduit or might be (SE cable)

service drop above roof

check roof/mast flashing here

12"

24"

0022

Using service entrance cable

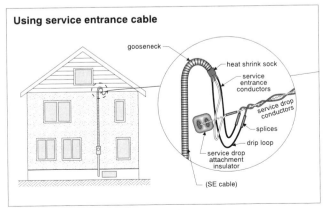

gooseneck

heat shrink sock

service entrance conductors

service drop conductors

splices

drip loop

service drop attachment insulator

(SE cable)

0023

Bent service masts

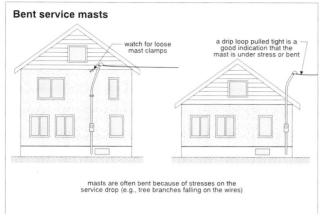

watch for loose mast clamps

a drip loop pulled tight is a good indication that the mast is under stress or bent

masts are often bent because of stresses on the service drop (e.g., tree branches falling on the wires)

0024

Support requirements for service entrances

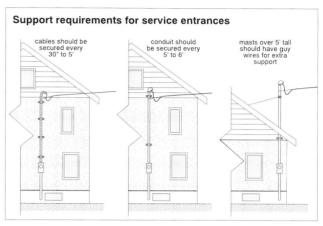

cables should be secured every 30" to 5'

conduit should be secured every 5' to 6'

masts over 5' tall should have guy wires for extra support

0025

Service entrance - areas of potential water entry

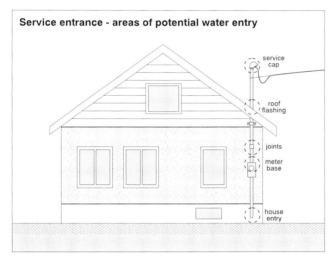

service cap

roof flashing

joints

meter base

house entry

0026

Poor seal at house/wall penetration

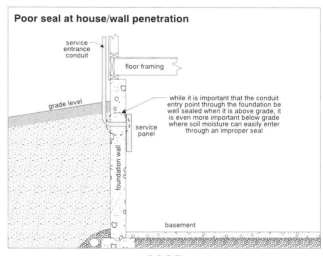

service entrance conduit

floor framing

grade level

while it is important that the conduit entry point through the foundation be well sealed when it is above grade, it is even more important below grade where soil moisture can easily enter through an improper seal

service panel

foundation wall

basement

0027

Service entrance cable should not be covered by siding

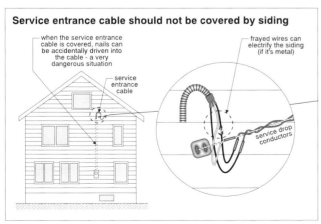

when the service entrance cable is covered, nails can be accidentally driven into the cable - a very dangerous situation

frayed wires can electrify the siding (if it's metal)

service entrance cable

service drop conductors

0028

Common wire sizes

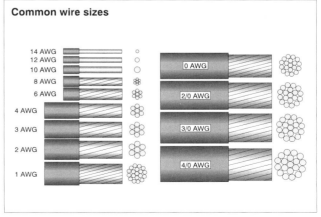

14 AWG
12 AWG
10 AWG
8 AWG
6 AWG
4 AWG
3 AWG
2 AWG
1 AWG

0 AWG
2/0 AWG
3/0 AWG
4/0 AWG

0029

Determining service size by the service entrance wires

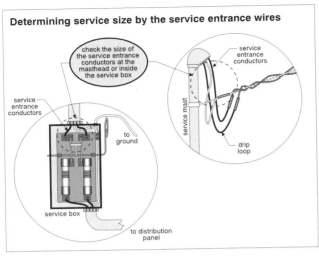

check the size of the service entrance conductors at the masthead or inside the service box

service entrance conductors

service entrance conductors

service mast

to ground

drip loop

service box

to distribution panel

0030

Determining service size by the main disconnect

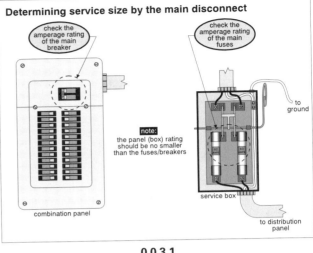

check the amperage rating of the main breaker

check the amperage rating of the main fuses

to ground

note: the panel (box) rating should be no smaller than the fuses/breakers

combination panel

service box

to distribution panel

0031

Service entrance (service box, main panel or service panel)

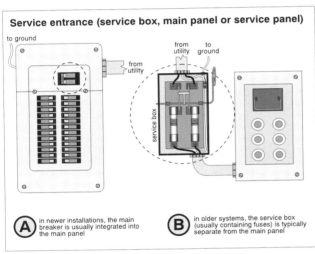

to ground

from utility

from utility   to ground

service box

A in newer installations, the main breaker is usually integrated into the main panel

B in older systems, the service box (usually containing fuses) is typically separate from the main panel

0032

Grounding equipment

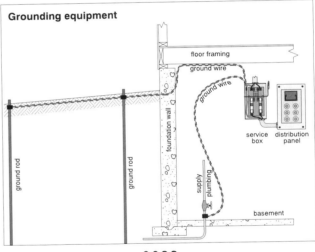

floor framing

ground wire

ground wire

foundation wall

ground rod

ground rod

supply plumbing

service box   distribution panel

basement

0033

Panels

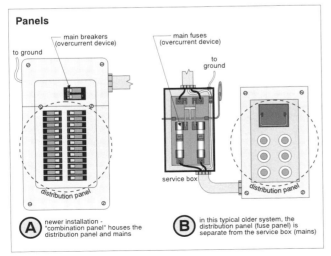

main breakers (overcurrent device)

main fuses (overcurrent device)

to ground

to ground

distribution panel

service box

distribution panel

distribution panel

A newer installation - "combination panel" houses the distribution panel and mains

B in this typical older system, the distribution panel (fuse panel) is separate from the service box (mains)

0034

Branch circuit overcurrent devices

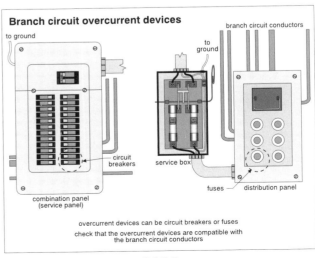

branch circuit conductors

to ground

to ground

circuit breakers

service box

fuses   distribution panel

combination panel (service panel)

overcurrent devices can be circuit breakers or fuses

check that the overcurrent devices are compatible with the branch circuit conductors

0035

Breaker and fuse type service boxes

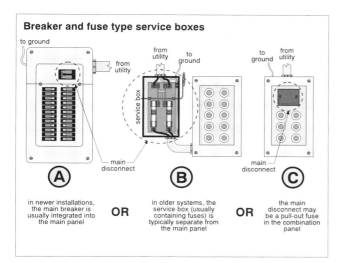

to ground

from utility

from utility

to ground

service box

main disconnect

A

in newer installations, the main breaker is usually integrated into the main panel

OR

in older systems, the service box (usually containing fuses) is typically separate from the main panel

OR

to ground   from utility

main disconnect

C

the main disconnect may be a pull-out fuse in the combination panel

B

0036

Meter location

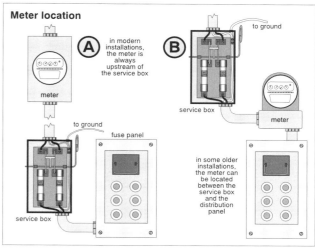

to ground

meter

A

in modern installations, the meter is always upstream of the service box

B

service box

meter

to ground   fuse panel

service box

in some older installations, the meter can be located between the service box and the distribution panel

0037

Panel clearances

3 feet of clearance should be provided in front of the panel

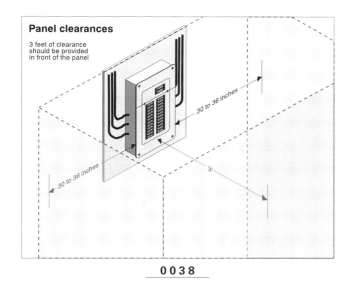

30 to 36 inches

30 to 36 inches

3'

0038

Securing the panel

the panel should be well secured to its support

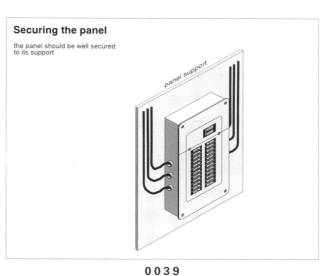

panel support

0039

Panel openings

any exposed panel openings (that would allow access to the inside of the panel) should be fitted with secure covers

panel support

openings requiring covers

0040

Panel mounting

in many jurisdictions, the panel support (or back-up) must be a non-combustible material such as drywall

plywood or wood planking were commonly used for panel support in older installations

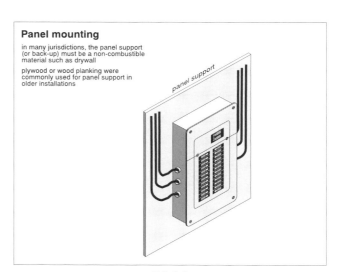

panel support

0041

Main fuses must be properly sized

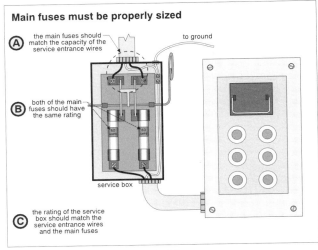

(A) the main fuses should match the capacity of the service entrance wires

to ground

(B) both of the main fuses should have the same rating

service box

(C) the rating of the service box should match the service entrance wires and the main fuses

0042

Illegal taps

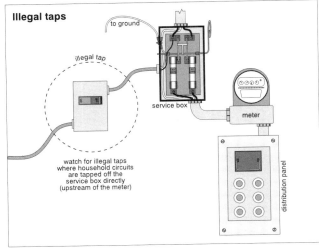

to ground

illegal tap

service box

meter

distribution panel

watch for illegal taps where household circuits are tapped off the service box directly (upstream of the meter)

0043

Neutral wire shouldn't bypass the service box

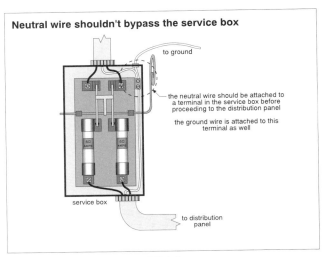

to ground

the neutral wire should be attached to a terminal in the service box before proceeding to the distribution panel

the ground wire is attached to this terminal as well

service box

to distribution panel

0044

The main fuses must be downstream of the disconnect switch

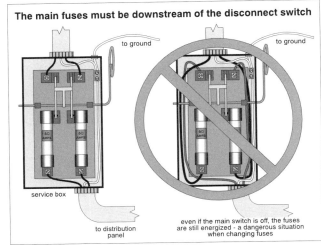

to ground

to ground

service box

to distribution panel

even if the main switch is off, the fuses are still energized - a dangerous situation when changing fuses

0045

Ground wires let fuses blow

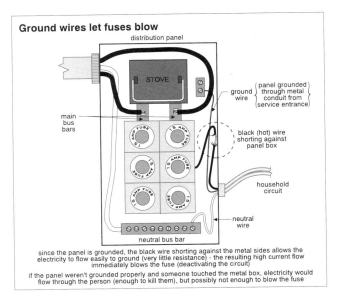

distribution panel

STOVE

ground wire {panel grounded through metal conduit from service entrance}

main bus bars

black (hot) wire shorting against panel box

household circuit

neutral wire

neutral bus bar

since the panel is grounded, the black wire shorting against the metal sides allows the electricity to flow easily to ground (very little resistance) - the resulting high current flow immediately blows the fuse (deactivating the circuit)

if the panel weren't grounded properly and someone touched the metal box, electricity would flow through the person (enough to kill them), but possibly not enough to blow the fuse

0046

Electrical path for ground and neutral wires

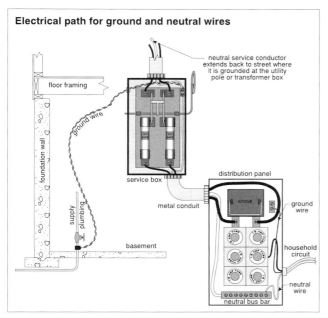

neutral service conductor extends back to street where it is grounded at the utility pole or transformer box

floor framing

ground wire

foundation wall

service box

metal conduit

supply plumbing

basement

distribution panel

STOVE

ground wire

household circuit

neutral wire

neutral bus bar

0047

Grounding the gas piping

in many areas, the gas piping must be bonded to the electrical grounding system (which typically means bonding to the supply piping)

supply plumbing

gas line

electrical service box

ground wire

foundation wall

basement

gas water heater

cross section

0048

Arcing and sparks

even though air is a good insulator, a spark will jump from a hot wire to a potential ground if the gap is small enough

this arcing can generate significant heat and lead to fires

black (hot) wire

ground wire

white (neutral) wire

nick in wire or excessive bending creates gap in wire

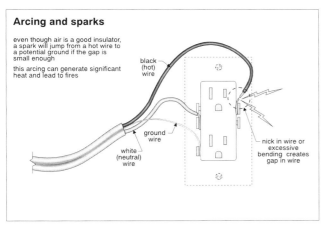

0049

Where do ground wires go?

ground rod   ground rod

grounding rods

metal grounding plates or rings

supply plumbing

metal water supply pipes

strip footing

buried in footings (UFER ground)

frames of metal buildings

metal well casing

metal casings of private wells

0050

Jumper wires

copper pipe   dielectric connector   galvanized pipe

(A) a jumper wire should be used to bridge around dielectric plumbing connectors

water meter

(B) a jumper wire should be installed around a water meter if the electrical service is grounded <u>downstream</u> of the meter

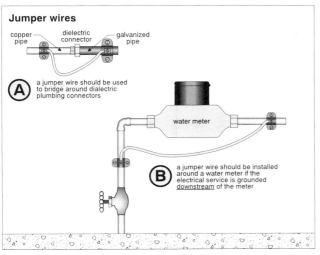

0051

Jumper wires needed

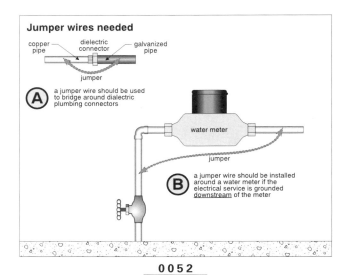

copper pipe — dielectric connector — galvanized pipe

jumper

A a jumper wire should be used to bridge around dialectric plumbing connectors

water meter

jumper

B a jumper wire should be installed around a water meter if the electrical service is grounded <u>downstream</u> of the meter

0052

Don't bond neutral and ground wires downstream of service box

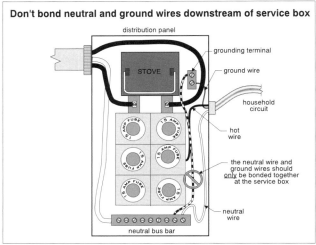

distribution panel

STOVE

grounding terminal

ground wire

household circuit

hot wire

the neutral wire and ground wires should <u>only</u> be bonded together at the service box

neutral wire

neutral bus bar

0053

Bond service box to ground

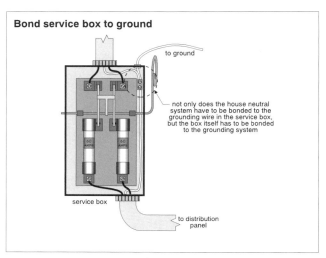

to ground

not only does the house neutral system have to be bonded to the grounding wire in the service box, but the box itself has to be bonded to the grounding system

service box

to distribution panel

0054

Need ground in subpanel feeder wires

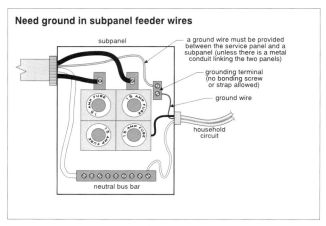

subpanel

a ground wire must be provided between the service panel and a subpanel (unless there is a metal conduit linking the two panels)

grounding terminal (no bonding screw or strap allowed)

ground wire

household circuit

neutral bus bar

0055

Typical arrangement of panel wires

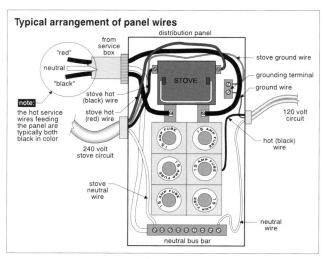

from service box

distribution panel

"red"

neutral

"black"

note: the hot service wires feeding the panel are typically both black in color

stove hot (black) wire

stove hot (red) wire

240 volt stove circuit

stove neutral wire

STOVE

stove ground wire

grounding terminal

ground wire

120 volt circuit

hot (black) wire

neutral wire

neutral bus bar

0056

Pull-out fuse blocks for 240 volt circuits

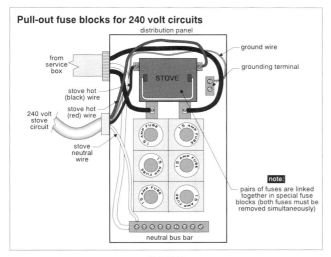

distribution panel

from service box

ground wire

grounding terminal

STOVE

stove hot (black) wire

stove hot (red) wire

240 volt stove circuit

stove neutral wire

note: pairs of fuses are linked together in special fuse blocks (both fuses must be removed simultaneously)

neutral bus bar

0057

Special circuit breakers for 240 volts

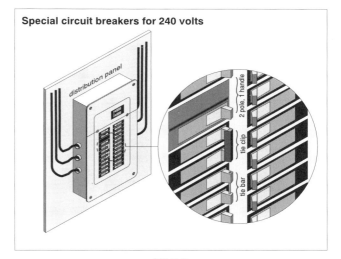

distribution panel

2 pole, 1 handle

tie clip

tie bar

0058

Common household wire and fuse sizes

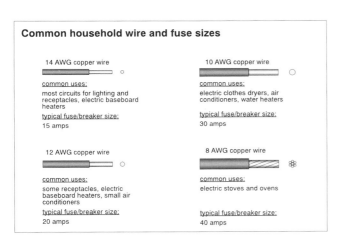

14 AWG copper wire

<u>common uses:</u>
most circuits for lighting and receptacles, electric baseboard heaters

<u>typical fuse/breaker size:</u>
15 amps

10 AWG copper wire

<u>common uses:</u>
electric clothes dryers, air conditioners, water heaters

<u>typical fuse/breaker size:</u>
30 amps

12 AWG copper wire

<u>common uses:</u>
some receptacles, electric baseboard heaters, small air conditioners

<u>typical fuse/breaker size:</u>
20 amps

8 AWG copper wire

<u>common uses:</u>
electric stoves and ovens

<u>typical fuse/breaker size:</u>
40 amps

0059

Inspecting the service box and panels

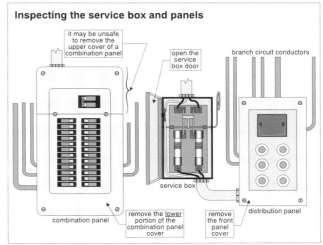

it may be unsafe to remove the upper cover of a combination panel

open the service box door

branch circuit conductors

service box

combination panel

remove the <u>lower</u> portion of the combination panel cover

remove the front panel cover

distribution panel

0060

Fuse types

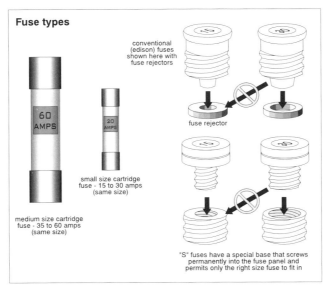

conventional (edison) fuses shown here with fuse rejectors

fuse rejector

60 AMPS

20 AMPS

small size cartridge fuse - 15 to 30 amps (same size)

medium size cartridge fuse - 35 to 60 amps (same size)

"S" fuses have a special base that screws permanently into the fuse panel and permits only the right size fuse to fit in

0061

Fused neutrals in old wiring systems

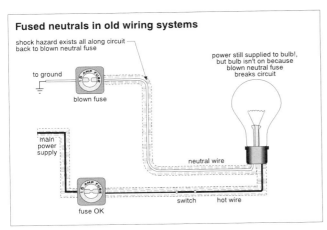

shock hazard exists all along circuit back to blown neutral fuse

to ground

blown fuse

power still supplied to bulb!, but bulb isn't on because blown neutral fuse breaks circuit

main power supply

fuse OK

neutral wire

switch   hot wire

0062

Subpanel wiring - no separate disconnect switch

a very poor arrangement - the feed wire is not protected against overload or short circuit

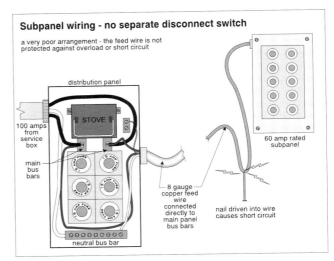

distribution panel

STOVE

100 amps from service box

main bus bars

neutral bus bar

60 amp rated subpanel

8 gauge copper feed wire connected directly to main panel bus bars

nail driven into wire causes short circuit

0063

Subpanel wiring - disconnect by subpanel

this is not good practice - although the fused disconnect protects the feed wire from <u>overloading</u>, a short circuit (as shown) could be enough to overheat the feed wire without either the disconnect fuses or the main fuses "noticing"

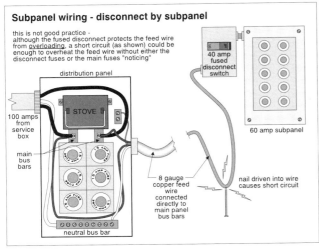

distribution panel

STOVE

100 amps from service box

main bus bars

neutral bus bar

40 amp fused disconnect switch

60 amp subpanel

8 gauge copper feed wire connected directly to main panel bus bars

nail driven into wire causes short circuit

0064

Subpanel wiring - disconnect by main panel

this is a better arrangement - the feed wire is protected against both overload and short circuit

best practice would be to move the stove circuit to the subpanel and use the existing stove fuses for the subpanel (this would eliminate the double-lugging)

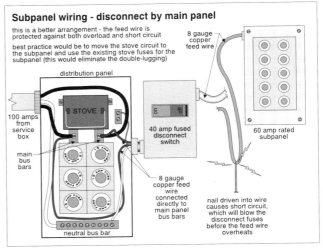

8 gauge copper feed wire

distribution panel

STOVE

100 amps from service box

main bus bars

neutral bus bar

40 amp fused disconnect switch

60 amp rated subpanel

8 gauge copper feed wire connected directly to main panel bus bars

nail driven into wire causes short circuit, which will blow the disconnect fuses before the feed wire overheats

0065

Double tapping (double lugging)

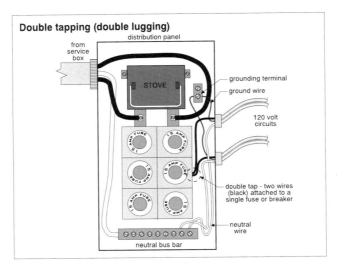

distribution panel

from service box

STOVE

grounding terminal

ground wire

120 volt circuits

double tap - two wires (black) attached to a single fuse or breaker

neutral wire

neutral bus bar

0066

Pigtailing to avoid double taps

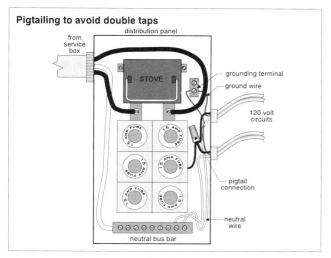

distribution panel

from service box

STOVE

grounding terminal

ground wire

120 volt circuits

pigtail connection

neutral wire

neutral bus bar

0067

Overcurrent protection for multi-wire branch circuits

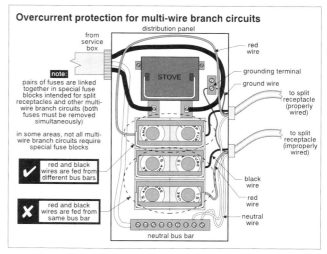

distribution panel

from service box

STOVE

red wire

grounding terminal

ground wire

to split receptacle (properly wired)

to split receptacle (improperly wired)

note: pairs of fuses are linked together in special fuse blocks intended for split receptacles and other multi-wire branch circuits (both fuses must be removed simultaneously)

in some areas, not all multi-wire branch circuits require special fuse blocks

✓ red and black wires are fed from different bus bars

✗ red and black wires are fed from same bus bar

black wire

red wire

neutral wire

neutral bus bar

0068

Staggered bus bars on circuit breaker panels

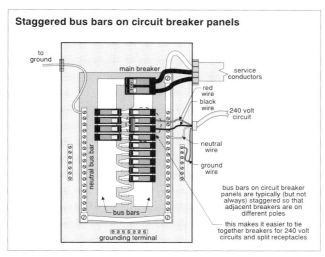

to ground

main breaker

service conductors

red wire

black wire

240 volt circuit

neutral wire

ground wire

neutral bus bar

bus bars on circuit breaker panels are typically (but not always) staggered so that adjacent breakers are on different poles

this makes it easier to tie together breakers for 240 volt circuits and split receptacles

bus bars

grounding terminal

0069

Bus bars in fuse panels

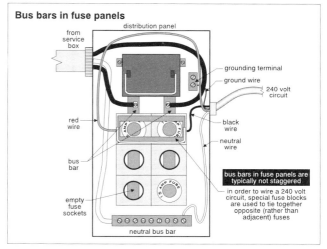

distribution panel

from service box

grounding terminal

ground wire

240 volt circuit

red wire

black wire

neutral wire

bus bar

bus bars in fuse panels are typically not staggered

empty fuse sockets

in order to wire a 240 volt circuit, special fuse blocks are used to tie together opposite (rather than adjacent) fuses

neutral bus bar

0070

Pull-out fuse blocks for multi-wire branch circuits and 240-volt circuits

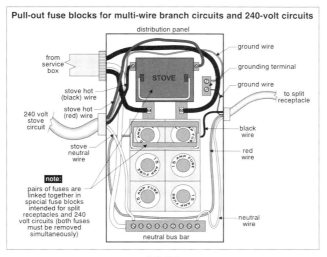

distribution panel

ground wire

from service box

STOVE

grounding terminal

ground wire

to split receptacle

stove hot (black) wire

stove hot (red) wire

240 volt stove circuit

black wire

red wire

stove neutral wire

note: pairs of fuses are linked together in special fuse blocks intended for split receptacles and 240 volt circuits (both fuses must be removed simultaneously)

neutral wire

neutral bus bar

0071

Excess sheathing on panel wires

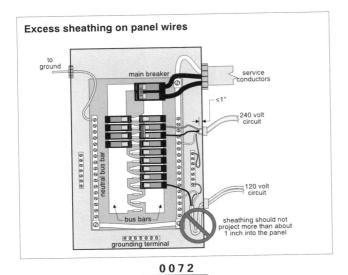

to ground

main breaker

service conductors

≤1"

240 volt circuit

120 volt circuit

sheathing should not project more than about 1 inch into the panel

neutral bus bar

bus bars

grounding terminal

0072

Wires shouldn't cross bus bars

distribution panel

from service box

STOVE

hot (black) wire

hot (red) wire

240 volt stove circuit

wires should not cross bus bars - they should be run <u>around</u> them

ground wire

grounding terminal

neutral wire

neutral bus bar

0073

Stranded wire for overhead runs

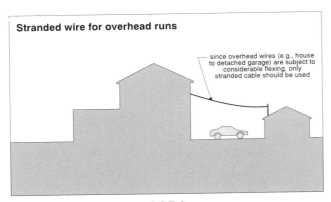

since overhead wires (e.g., house to detached garage) are subject to considerable flexing, only stranded cable should be used

0074

Number of conductors

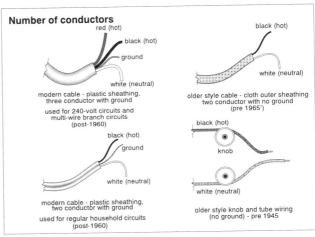

red (hot)
black (hot)
ground
white (neutral)

modern cable - plastic sheathing, three conductor with ground
used for 240-volt circuits and multi-wire branch circuits (post-1960)

black (hot)
ground
white (neutral)

modern cable - plastic sheathing, two conductor with ground
used for regular household circuits (post-1960)

black (hot)
white (neutral)

older style cable - cloth outer sheathing two conductor with no ground (pre 1965')

black (hot)
knob
white (neutral)

older style knob and tube wiring (no ground) - pre 1945

0075

Types of connections

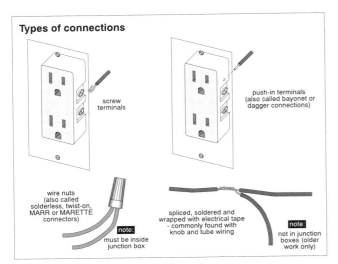

screw terminals

push-in terminals (also called bayonet or dagger connections)

wire nuts (also called solderless, twist-on, MARR or MARETTE connectors)

note: must be inside junction box

spliced, soldered and wrapped with electrical tape - commonly found with knob and tube wiring

note: not in junction boxes (older work only)

0076

Cable support inside walls

staples not required where cables run through holes in framing members

staple every 4-1/2 feet (USA)

staple every 5 feet (CANADA)

staple within 12 inches of electrical boxes

staple where cables change direction

0077

Wire support with steel studs

steel studs are very sharp

in order to protect the wire from damage, plastic grommets are required where wires pass through the studs

wires running along the studs should have plastic or wood standoffs to separate the wire from the metal by at least 1/4"

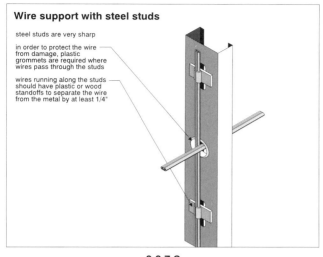

0078

Wire clearances from hot ducts and pipes

radiator piping

heating duct

single wall exhaust flue (wood)

single wall exhaust flue (oil)

single wall exhaust flue (gas)

wire

1"

1"

a piece of insulation can be used to separate the duct or pipe from the wire

18"

9" to 18"

6"

wire

wire

wire

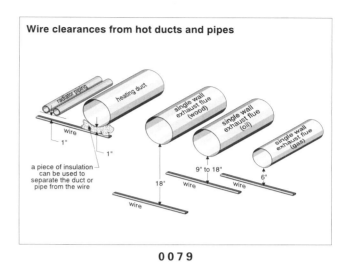

0079

Edge clearance for wires in studs and joists

protective metal plate is required if wire is within 1-1/4" of stud face

1-1/4" of clearance required

stud

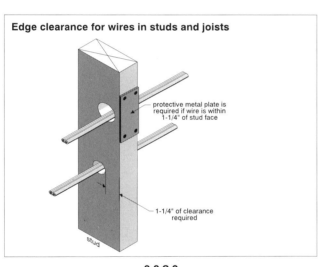

0080

Exposed wiring

exposed wires running along walls within 5 to 7 feet of the floor require protection from mechanical damage

wire doesn't require protection

wire requires protection

5' (CANADA)
7' (USA)

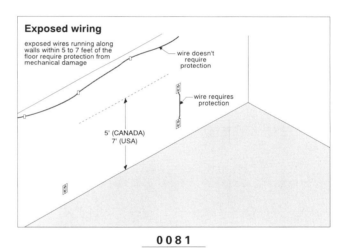

0081

Exposed wires in attics

wires should not be run on top of exposed attic rafters or joists

exception:
in some jurisdictions, this is permitted if headroom is less than 40"

roof rafter

40"

wires

wires

wires

ceiling joist

wall

cross section

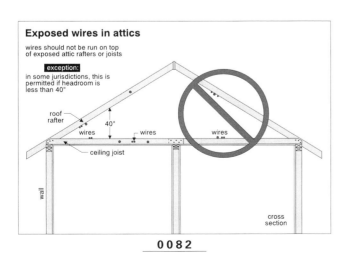

0082

Knob and tube wiring

hot

joist

neutral

ceramic tube

ceramic knob

perspective

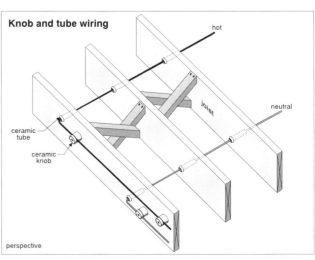

0083

Knob and tube connections

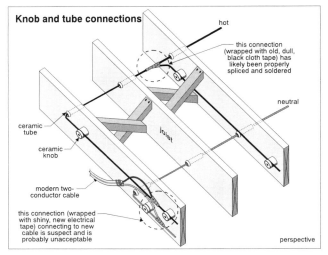

hot

this connection (wrapped with old, dull, black cloth tape) has likely been properly spliced and soldered

neutral

ceramic tube

ceramic knob

joist

modern two-conductor cable

this connection (wrapped with shiny, new electrical tape) connecting to new cable is suspect and is probably unacceptable

perspective

0 0 8 4

Knob and tube wiring - extending a circuit

two junction boxes (with a connecting wire between them) are required to properly tap into existing knob and tube wiring because there isn't usually enough slack in the old wire to make the all of the connections in one box

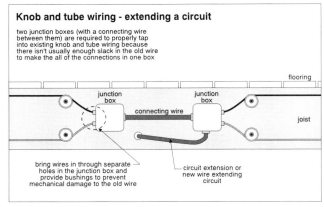

flooring

junction box

junction box

connecting wire

joist

bring wires in through separate holes in the junction box and provide bushings to prevent mechanical damage to the old wire

circuit extension or new wire extending circuit

0 0 8 5

CUAL designation

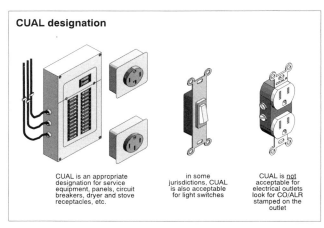

CUAL is an appropriate designation for service equipment, panels, circuit breakers, dryer and stove receptacles, etc.

in some jurisdictions, CUAL is also acceptable for light switches

CUAL is not acceptable for electrical outlets look for CO/ALR stamped on the outlet

0 0 8 6

Stairway lighting

stairway lighting requires switches at both the top and bottom of the stairs when the stairs have more than 3 treads (CAN) or more than 6 treads (USA)

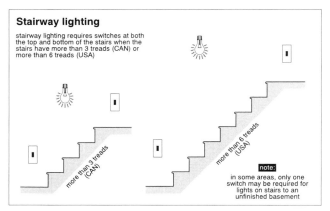

more than 3 treads (CAN)

more than 6 treads (USA)

note:
in some areas, only one switch may be required for lights on stairs to an unfinished basement

0 0 8 7

How three-way switches work

(simplified schematic)

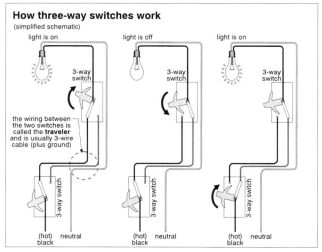

light is on

light is off

light is on

3-way switch

3-way switch

3-way switch

the wiring between the two switches is called the **traveler** and is usually 3-wire cable (plus ground)

3-way switch

3-way switch

3-way switch

(hot) black   neutral

(hot) black   neutral

(hot) black   neutral

0 0 8 8

Potlights in insulated ceilings

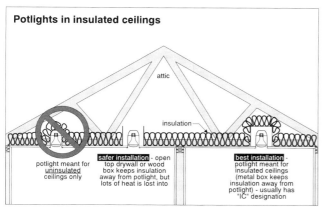

attic

insulation

potlight meant for <u>uninsulated</u> ceilings only

safer installation - open top drywall or wood box keeps insulation away from potlight, but lots of heat is lost into

best installation - potlight meant for insulated ceilings (metal box keeps insulation away from potlight) - usually has "IC" designation

__0089__

Use proper bulbs for potlights

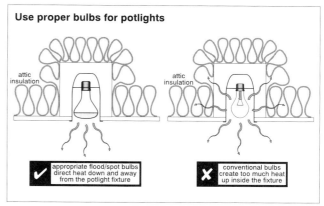

attic insulation

attic insulation

✓ appropriate flood/spot bulbs direct heat down and away from the potlight fixture

✗ conventional bulbs create too much heat up inside the fixture

__0090__

Poor heat lamp locations

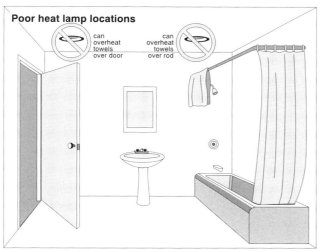

can overheat towels over door

can overheat towels over rod

__0091__

Isolating links needed on pull chains

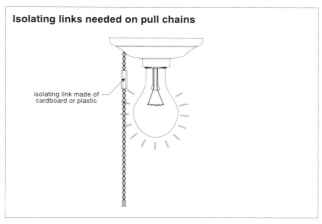

isolating link made of cardboard or plastic

__0092__

Reversed polarity

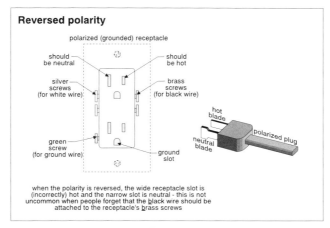

polarized (grounded) receptacle

should be neutral

should be hot

silver screws (for white wire)

brass screws (for black wire)

green screw (for ground wire)

ground slot

hot blade

neutral blade

polarized plug

when the polarity is reversed, the wide receptacle slot is (incorrectly) hot and the narrow slot is neutral - this is not uncommon when people forget that the <u>b</u>lack wire should be attached to the receptacle's <u>b</u>rass screws

__0093__

Importance of correct polarity with light fixtures

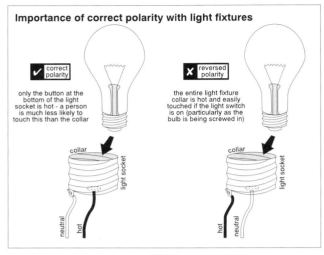

✔ correct polarity

only the button at the bottom of the light socket is hot - a person is much less likely to touch this than the collar

✘ reversed polarity

the entire light fixture collar is hot and easily touched if the light switch is on (particularly as the bulb is being screwed in)

collar

light socket

neutral   hot

collar

light socket

hot   neutral

0094

Wrong type receptacle

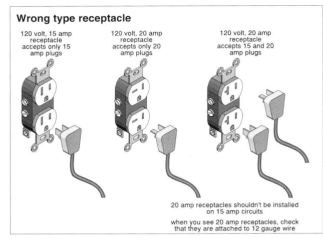

120 volt, 15 amp receptacle accepts only 15 amp plugs

120 volt, 20 amp receptacle accepts only 20 amp plugs

120 volt, 20 amp receptacle accepts 15 and 20 amp plugs

20 amp receptacles shouldn't be installed on 15 amp circuits

when you see 20 amp receptacles, check that they are attached to 12 gauge wire

0095

Ground fault interrupter

the GFI circuitry within the outlet checks constantly for a difference between the current in the black and white wires

if there is a difference (even as little as 5 milliamps), there is a current leak (possibly through your body) and the GFI shuts down the receptacle and other receptacles downstream

note:
if the GFI is in the panel, the entire circuit will be shut down

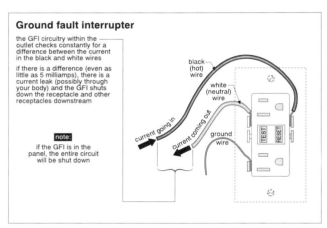

black (hot) wire

white (neutral) wire

ground wire

current going in

current coming out

TEST   RESET

0096

GFIs can protect ordinary outlets downstream

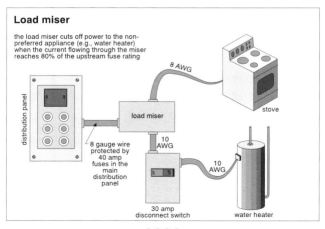

master bedroom bathroom

main floor washroom GFI

builders often use one GFI (often in a mainfloor washroom) to protect all of the bathroom outlets in the house

basement bathroom

main second floor bathroom

0097

Load miser

the load miser cuts off power to the non-preferred appliance (e.g., water heater) when the current flowing through the miser reaches 80% of the upstream fuse rating

8 AWG

load miser

stove

distribution panel

8 gauge wire protected by 40 amp fuses in the main distribution panel

10 AWG

10 AWG

30 amp disconnect switch

water heater

0098

Bathroom outlets and switches

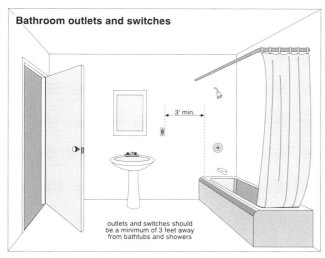

3' min.

outlets and switches should
be a minimum of 3 feet away
from bathtubs and showers

0099

Outlets near basins

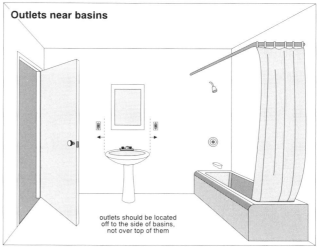

outlets should be located
off to the side of basins,
not over top of them

0100

Outlets should not be above electric baseboard heaters

cords plugged into outlets above electric
baseboard heaters could overheat if
accidentally draped over the heater

outlets should be located at either <u>end</u> of the
heater

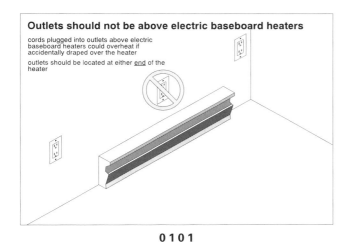

0101

Outlets in floors or countertops

in general, electrical
outlets should not be
flush-mounted on
horizontal surfaces

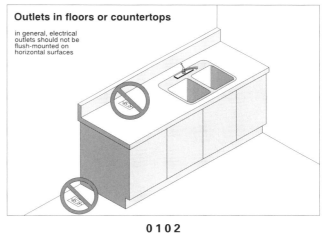

0102

Garage outlets should be at least 18 inches off the floor

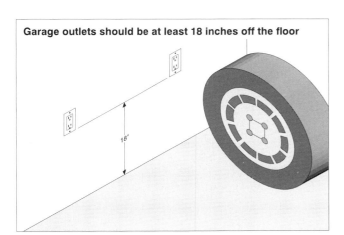

18"

0103

Old pushbutton switches are obsolete

(but watch for reproductions that are acceptable)

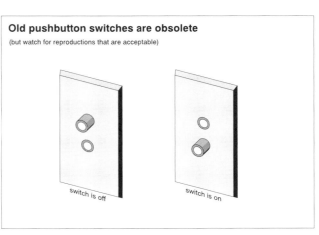

switch is off   switch is on

0104

Furnace switches

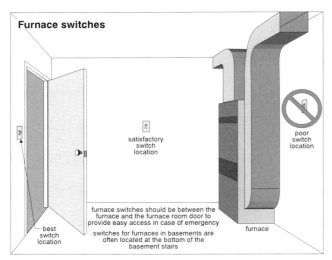

satisfactory
switch
location

poor
switch
location

furnace switches should be between the
furnace and the furnace room door to
provide easy access in case of emergency

switches for furnaces in basements are
often located at the bottom of the
basement stairs

best
switch
location

furnace

0105

Garbage disposal switches

the garbage disposal switch is best
mounted above the counter or in the
cupboard below the sink rather than
on the front face of the base cabinets
where it is prone to damage

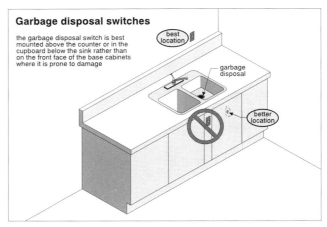

best
location

garbage
disposal

better
location

0106

Garbage disposal wiring

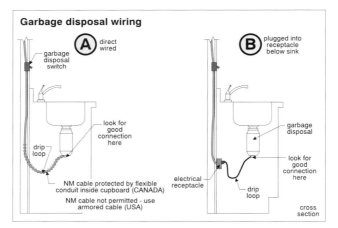

A direct
wired

garbage
disposal
switch

look for
good
connection
here

drip
loop

NM cable protected by flexible
conduit inside cupboard (CANADA)

NM cable not permitted - use
armored cable (USA)

B plugged into
receptacle
below sink

garbage
disposal

look for
good
connection
here

electrical
receptacle

drip
loop

cross
section

0107

PART 2

PLUMBING

Factors affecting supply of water

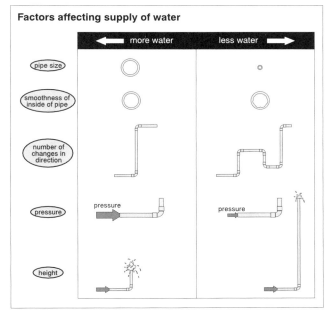

0108

Pressure drop as flow rate increases

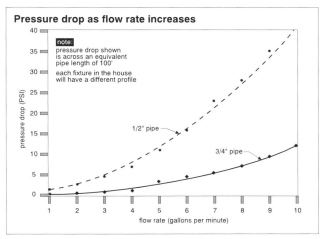

note:
pressure drop shown
is across an equivalent
pipe length of 100'

each fixture in the house
will have a different profile

1/2" pipe

3/4" pipe

pressure drop (PSI)

flow rate (gallons per minute)

0109

3/4" pipe is more than twice as big as 1/2" pipe!

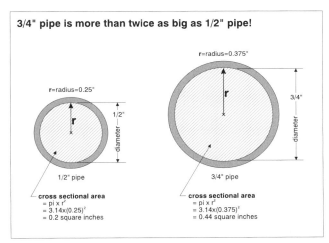

0110

Galvanized steel pipe

rusting of galvanized pipe can greatly reduce water
pressure and will eventually cause leaks as rust creates
holes in the pipe walls

problems are likely to occur soonest on pipes carrying hot
water, horizontal pipes and at threaded (thinner) sections

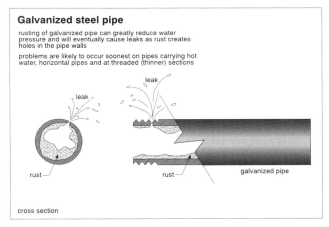

0111

Downstream versus upstream pipe replacement

installing larger diameter piping in the downstream sections
is just as effective as replacing the upstream sections

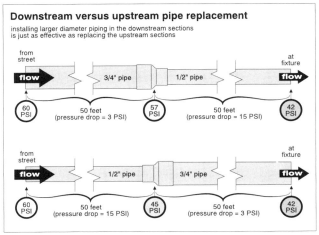

0112

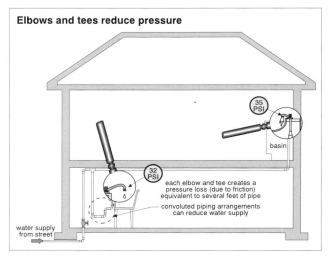

Elbows and tees reduce pressure

35 PSI

basin

32 PSI

each elbow and tee creates a pressure loss (due to friction) equivalent to several feet of pipe

convoluted piping arrangements can reduce water supply

water supply from street

0113

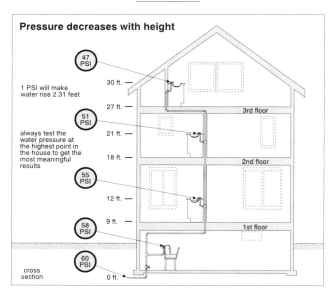

Pressure regulator required

residential systems with static pressure in excess of 80 PSI usually have a pressure regulator

foundation wall

pressure relief valve may be required (typical setting - 150 PSI)

pressure regulator (maximum 80 PSI)

water meter

main water shut off valve

pressure relief discharge pipe

basement

footing

supply plumbing from street

a strainer should be installed upstream (these are often part of the regulator)

floor drain

0114

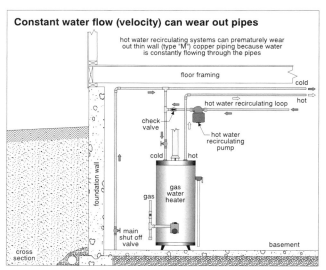

Constant water flow (velocity) can wear out pipes

hot water recirculating systems can prematurely wear out thin wall (type "M") copper piping because water is constantly flowing through the pipes

floor framing

cold

hot water recirculating loop

hot

check valve

hot water recirculating pump

cold   hot

foundation wall

gas

gas water heater

main shut off valve

basement

cross section

0115

Pressure decreases with height

47 PSI

30 ft.

1 PSI will make water rise 2.31 feet

27 ft.

3rd floor

51 PSI

21 ft.

always test the water pressure at the highest point in the house to get the most meaningful results

18 ft.

2nd floor

55 PSI

12 ft.

58 PSI

9 ft.

1st floor

cross section

60 PSI

0 ft.

0116

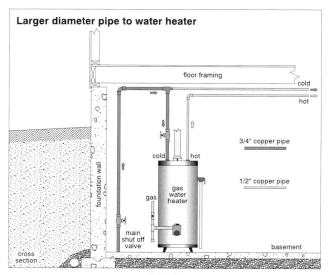

Larger diameter pipe to water heater

floor framing

cold

hot

cold   hot

3/4" copper pipe

foundation wall

1/2" copper pipe

gas

gas water heater

main shut off valve

basement

cross section

0117

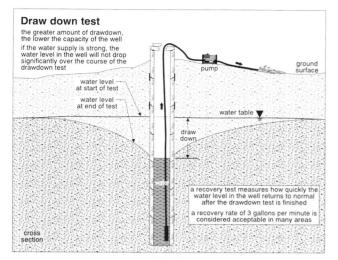

Draw down test

the greater amount of drawdown, the lower the capacity of the well

if the water supply is strong, the water level in the well will not drop significantly over the course of the drawdown test

pump

ground surface

water level at start of test

water level at end of test

water table

draw down

a recovery test measures how quickly the water level in the well returns to normal after the drawdown test is finished

a recovery rate of 3 gallons per minute is considered acceptable in many areas

water

cross section

0118

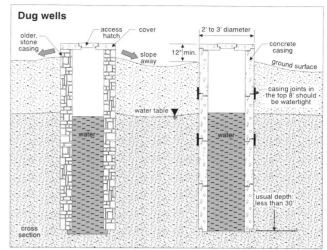

Dug wells

older, stone casing

access hatch

cover

2' to 3' diameter

concrete casing

slope away

12" min.

ground surface

water table

casing joints in the top 8' should be watertight

water

water

usual depth: less than 30'

cross section

0119

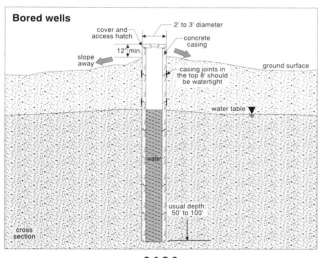

Bored wells

cover and access hatch

2' to 3' diameter

concrete casing

slope away

12" min.

casing joints in the top 8' should be watertight

ground surface

water table

water

usual depth: 50' to 100'

cross section

0120

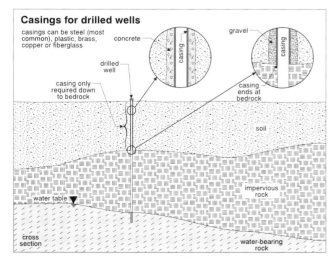

Drilled wells

casing only required down to bedrock

4" to 6" diameter and 50' to 900" deep

ground surface

concrete

soil

casing

impervious rock

water table

casing

gravel

casing ends at bedrock

water-bearing rock

cross section

0121

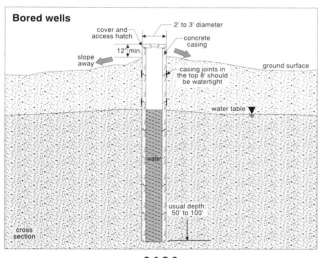

Casings for drilled wells

casings can be steel (most common), plastic, brass, copper or fiberglass

concrete

gravel

casing

casing

drilled well

casing only required down to bedrock

casing ends at bedrock

soil

water table

impervious rock

cross section

water-bearing rock

0122

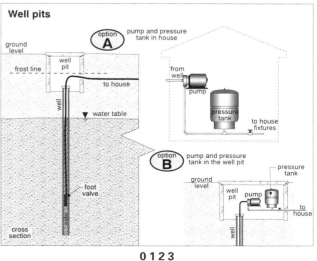

Well pits

option A

pump and pressure tank in house

ground level

frost line

well pit

to house

well

water table

foot valve

from well

pump

pressure tank

to house fixtures

option B

pump and pressure tank in the well pit

ground level

pressure tank

well pit

pump

to house

well

cross section

0123

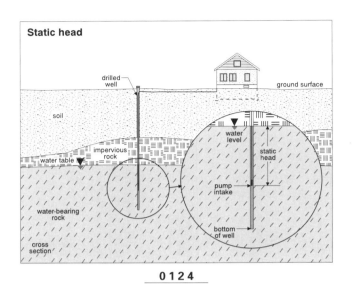

Static head

drilled well

ground surface

soil

water level

static head

impervious rock

water table

pump intake

water-bearing rock

bottom of well

cross section

0124

Well location

tile bed

house

septic tank

50' minimum

50' minimum if well has a casing
100' minimum if well has <u>no</u> casing

well

plan view

0125

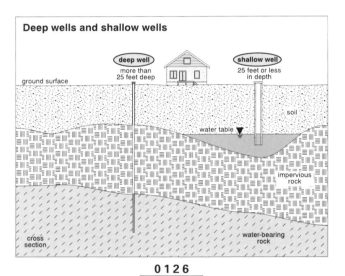

Deep wells and shallow wells

deep well
more than 25 feet deep

shallow well
25 feet or less in depth

ground surface

soil

water table

impervious rock

cross section

water-bearing rock

0126

Foot valve

well

suction line

ground surface

pump

pump operating

pump off
water stays in suction line

suction opens foot valve

strainer

spring closes foot valve

water

pump intake (and foot valve)

cross section

0127

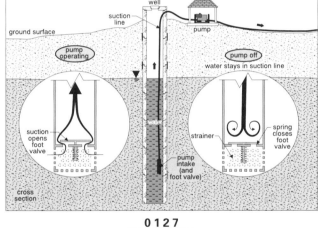

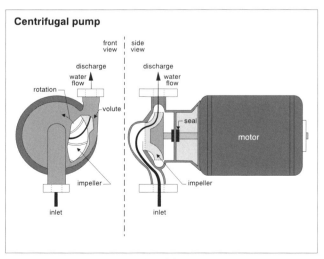

Types of pumps

water to house

2-line jet pump (jet is in well)

from well

to well

water to house

from well

1-line jet pump (jet is at pump)

1-line jet pump (shallow well)

jet in well

water from pump

well water sucked up by venturi action

casing/hole

2-line jet pump (deep well)

pump

casing/hole

submersible pump (deep well)
most powerful

0128

Centrifugal pump

front view

side view

discharge

water flow

rotation

volute

discharge

water flow

seal

motor

impeller

impeller

inlet

inlet

0129

Shallow well with single line jet pump

this arrangement is suitable for
shallow wells less than 27 feet deep

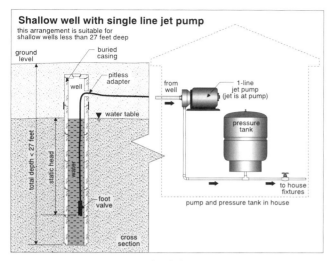

0130

Two-line jet pump

suitable for deeper wells

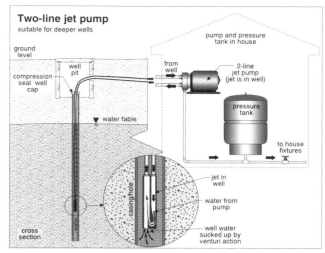

0131

Piston pump

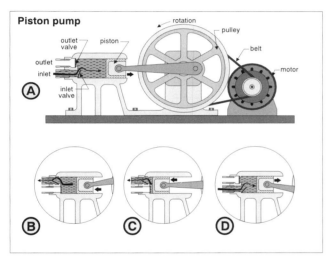

0132

Drilled well with submersible pump

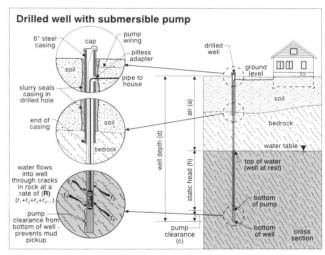

0133

Pressure tank components and pump controls

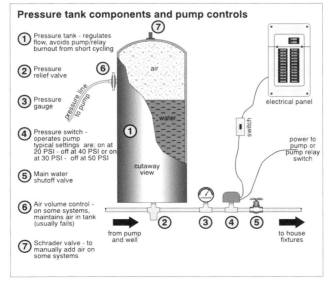

(1) Pressure tank - regulates flow, avoids pump/relay burnout from short cycling

(2) Pressure relief valve

(3) Pressure gauge

(4) Pressure switch - operates pump typical settings are: on at 20 PSI - off at 40 PSI or on at 30 PSI - off at 50 PSI

(5) Main water shutoff valve

(6) Air volume control - on some systems, maintains air in tank (usually fails)

(7) Schrader valve - to manually add air on some systems

0134

Bladder-type pressure tank

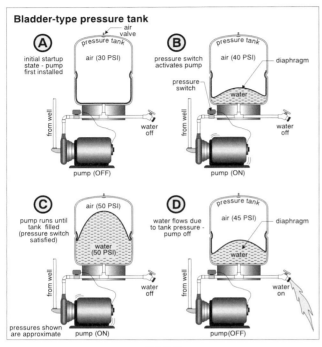

A initial startup state - pump first installed

pressure tank

air (30 PSI)

air valve

from well

water off

pump (OFF)

B pressure switch activates pump

pressure tank

air (40 PSI)

diaphragm

pressure switch

water

from well

water off

pump (ON)

C pump runs until tank filled (pressure switch satisfied)

air (50 PSI)

water (50 PSI)

from well

water off

pressures shown are approximate   pump (ON)

D water flows due to tank pressure - pump off

pressure tank

air (45 PSI)

diaphragm

water

from well

water on

pump(OFF)

0135

Waterlogged pressure tank - short-cycling

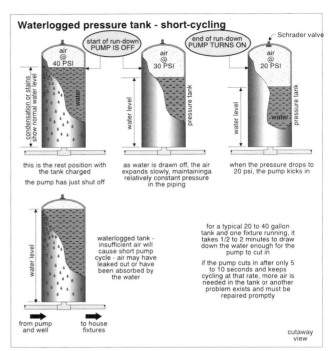

Schrader valve

start of run-down PUMP IS OFF

end of run-down PUMP TURNS ON

air @ 40 PSI

air @ 30 PSI

air @ 20 PSI

condensation or stains show normal water level

water

water level

pressure tank

water level

water

pressure tank

this is the rest position with the tank charged

the pump has just shut off

as water is drawn off, the air expands slowly, maintaininga relatively constant pressure in the piping

when the pressure drops to 20 psi, the pump kicks in

water level

waterlogged tank - insufficient air will cause short pump cycle - air may have leaked out or have been absorbed by the water

from pump and well

to house fixtures

for a typical 20 to 40 gallon tank and one fixture running, it takes 1/2 to 2 minutes to draw down the water enough for the pump to cut in

if the pump cuts in after only 5 to 10 seconds and keeps cycling at that rate, more air is needed in the tank or another problem exists and must be repaired promptly

cutaway view

0136

Service piping - when different types were used

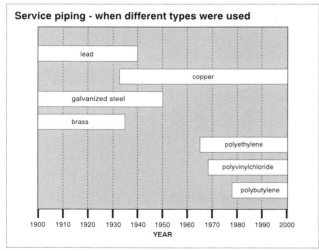

lead

copper

galvanized steel

brass

polyethylene

polyvinylchloride

polybutylene

1900 1910 1920 1930 1940 1950 1960 1970 1980 1990 2000

YEAR

0137

Separate water service from sewer pipe

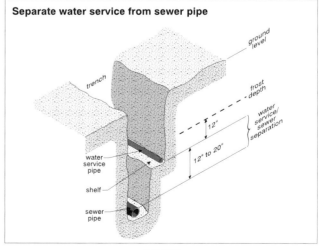

ground level

trench

frost depth

water service/ sewer separation

12"

12" to 20"

water service pipe

shelf

sewer pipe

0138

Main shut off valve - stop and waste

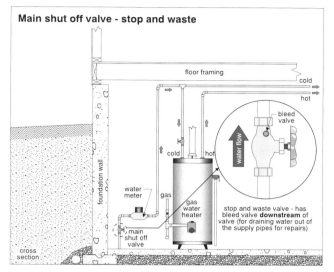

floor framing

cold

hot

bleed valve

water flow

cold

hot

foundation wall

water meter

gas

gas water heater

main shut off valve

cross section

stop and waste valve - has bleed valve **downstream** of valve (for draining water out of the supply pipes for repairs)

0139

Globe valve

closed

handle

packing nut

packing washer

spindle

water flow

supply pipe

stem washer

glove valves tend to be restrictive but, can be used to throttle water flow

valve seat

open

water flow

water flow

0140

Gate valve

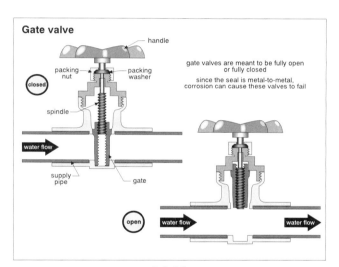

handle

closed

packing nut

packing washer

spindle

water flow

supply pipe

gate

gate valves are meant to be fully open or fully closed

since the seal is metal-to-metal, corrosion can cause these valves to fail

open

water flow

water flow

0141

Ball valve

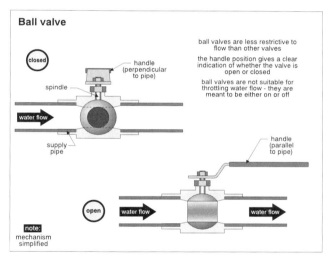

closed

handle (perpendicular to pipe)

spindle

water flow

supply pipe

ball valves are less restrictive to flow than other valves

the handle position gives a clear indication of whether the valve is open or closed

ball valves are not suitable for throttling water flow - they are meant to be either on or off

handle (parallel to pipe)

note: mechanism simplified

open

water flow

water flow

0142

Types of copper supply pipe

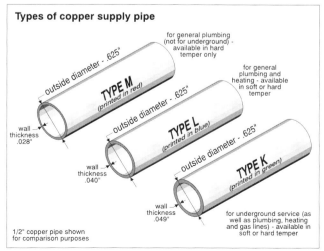

outside diameter - .625"

TYPE M (printed in red)

for general plumbing (not for underground) - available in hard temper only

wall thickness .028"

outside diameter - .625"

TYPE L (printed in blue)

for general plumbing and heating - available in soft or hard temper

wall thickness .040"

outside diameter - .625"

TYPE K (printed in green)

for underground service (as well as plumbing, heating and gas lines) - available in soft or hard temper

wall thickness .049"

1/2" copper pipe shown for comparison purposes

0143

Soldered pipe connection

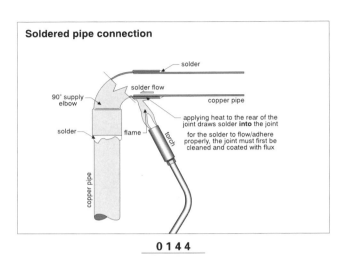

solder

solder flow

90° supply elbow

copper pipe

solder

flame

torch

applying heat to the rear of the joint draws solder **into** the joint

for the solder to flow/adhere properly, the joint must first be cleaned and coated with flux

copper pipe

0 1 4 4

Flare fitting

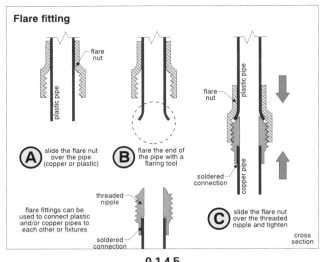

flare nut

plastic pipe

(A) slide the flare nut over the pipe (copper or plastic)

(B) flare the end of the pipe with a flaring tool

flare nut

plastic pipe

soldered connection

copper pipe

(C) slide the flare nut over the threaded nipple and tighten

threaded nipple

flare fittings can be used to connect plastic and/or copper pipes to each other or fixtures

soldered connection

cross section

0 1 4 5

Compression fitting

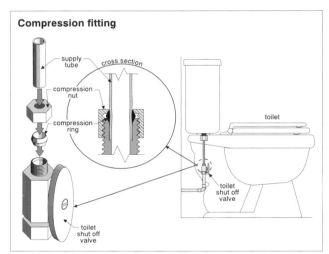

supply tube

cross section

compression nut

compression ring

toilet

toilet shut off valve

toilet shut off valve

0 1 4 6

Dielectric union

a dielectric union should be used when joining galvanized steel pipe to copper pipe - the insulating sleeve and washer separate the two dissimilar metals to prevent corrosion

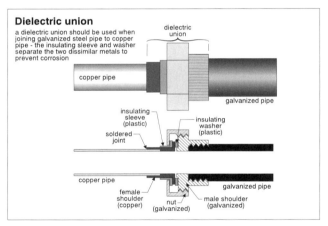

dielectric union

copper pipe

galvanized pipe

insulating sleeve (plastic)

insulating washer (plastic)

soldered joint

copper pipe

galvanized pipe

female shoulder (copper)

nut (galvanized)

male shoulder (galvanized)

0 1 4 7

Hangers for copper pipe

copper pipe should be supported on brass or copper hangers

it should not be supported on nails or uninsulated steel hangers (otherwise, localized deterioration of the pipe, or hanger, can occur)

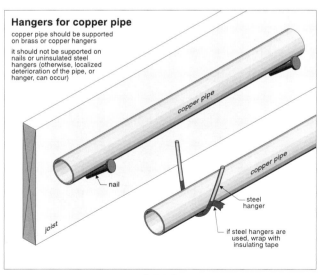

copper pipe

copper pipe

nail

joist

steel hanger

if steel hangers are used, wrap with insulating tape

0 1 4 8

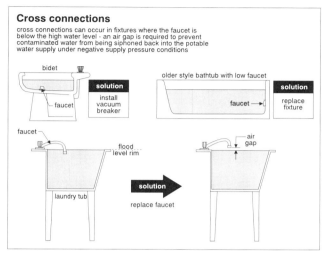

Cross connections

cross connections can occur in fixtures where the faucet is below the high water level - an air gap is required to prevent contaminated water from being siphoned back into the potable water supply under negative supply pressure conditions

bidet

solution
install vacuum breaker

faucet

older style bathtub with low faucet

solution
replace fixture

faucet

faucet

flood level rim

solution

laundry tub

replace faucet

air gap

0149

Air gap

air gap

high water level

counter top

water

kitchen sink

trap

0150

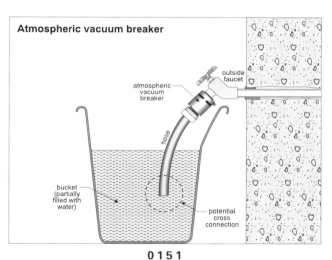

Atmospheric vacuum breaker

atmospheric vacuum breaker

outside faucet

hose

bucket (partially filled with water)

potential cross connection

0151

Pressure type vacuum breaker

valve open

normal operation

outlet

water flow

inlet

vacuum breaker

bidet

backflow condition

air

shut off valve

valve closed

0152

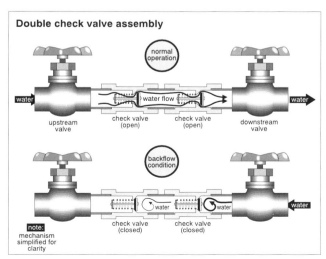

Double check valve assembly

water

normal operation

water flow

water

upstream valve

check valve (open)

check valve (open)

downstream valve

backflow condition

water

water

water

check valve (closed)

check valve (closed)

note:
mechanism simplified for clarity

0153

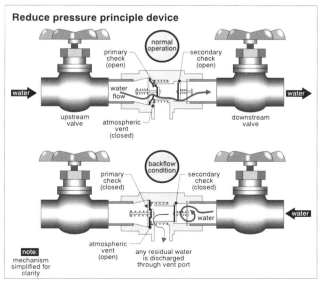

Reduce pressure principle device

normal operation

primary check (open)

secondary check (open)

water

water flow

water

upstream valve

atmospheric vent (closed)

downstream valve

backflow condition

primary check (closed)

secondary check (closed)

water

atmospheric vent (open)

any residual water is discharged through vent port

water

note:
mechanism simplified for clarity

0154

Dishwasher air gap

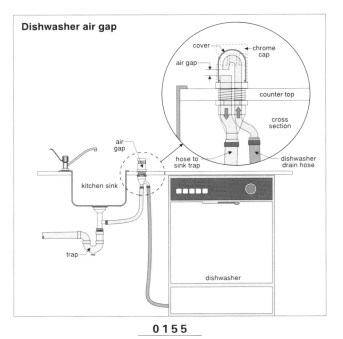

0155

Correcting water hammer

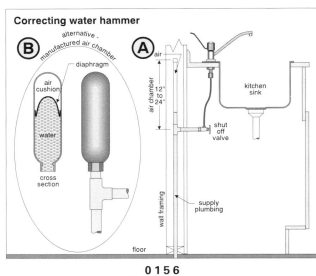

0156

Vent connectors and clearances to plastic supply pipe

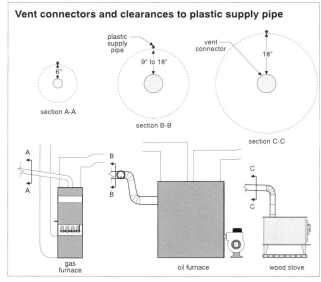

0157

Polybutylene pipe should be at least 18" from water heaters

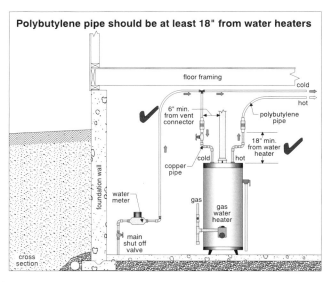

0158

Polybutylene pipe - crimp fitting

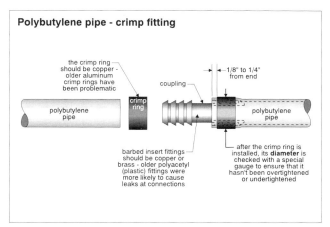

the crimp ring should be copper - older aluminum crimp rings have been problematic

coupling

1/8" to 1/4" from end

polybutylene pipe

crimp ring

polybutylene pipe

barbed insert fittings should be copper or brass - older polyacetyl (plastic) fittings were more likely to cause leaks at connections

after the crimp ring is installed, its **diameter** is checked with a special gauge to ensure that it hasn't been overtightened or undertightened

0159

Polybutylene pipe - compression (grip) fitting

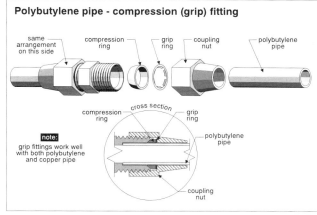

same arrangement on this side

compression ring

grip ring

coupling nut

polybutylene pipe

compression ring

cross section

grip ring

polybutylene pipe

note: grip fittings work well with both polybutylene and copper pipe

coupling nut

0160

Bend radius for polybutylene pipe

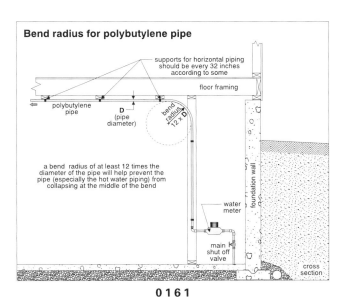

supports for horizontal piping should be every 32 inches according to some

floor framing

polybutylene pipe

D (pipe diameter)

bend radius 12 x **D**

a bend radius of at least 12 times the diameter of the pipe will help prevent the pipe (especially the hot water piping) from collapsing at the middle of the bend

water meter

main shut off valve

foundation wall

cross section

0161

Sacrificial anode

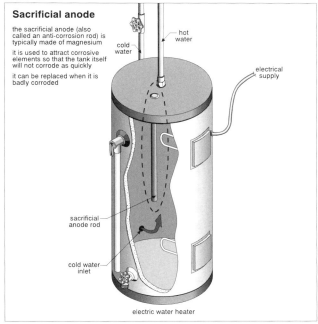

the sacrificial anode (also called an anti-corrosion rod) is typically made of magnesium

it is used to attract corrosive elements so that the tank itself will not corrode as quickly

it can be replaced when it is badly corroded

cold water

hot water

electrical supply

sacrificial anode rod

cold water inlet

electric water heater

0162

Vacuum relief valve

vacuum relief valves are installed in some jurisdictions to
prevent hot water from backing into the cold water lines
in the event of a cold water pressure drop

they also protect against possible collapse of the storage
tank by preventing a vacuum from forming inside

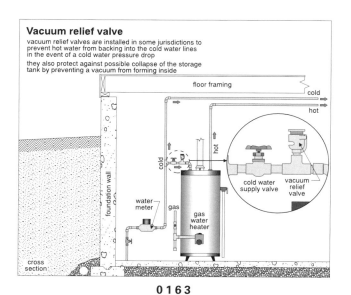

floor framing

cold

hot

foundation wall

cold

hot

cold water
supply valve

vacuum
relief
valve

water
meter

gas

gas
water
heater

cross
section

0 1 6 3

Oil-fired water heater

cold
water
supply
valve

cold
water

hot
water

vent
connector

insulation

dip
tube

oil
burner

sacrificial
anode rod

temperature/
pressure
relief valve

discharge
pipe

baffle
(turbulator)

flue

drain
valve

refractory

0 1 6 4

Gas water heater

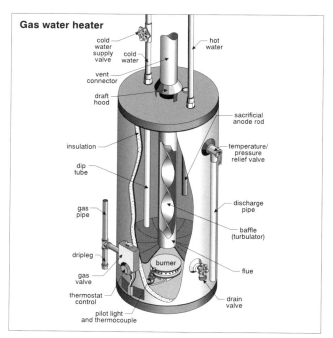

cold
water
supply
valve

cold
water

hot
water

vent
connector

draft
hood

insulation

dip
tube

gas
pipe

dripleg

gas
valve

thermostat
control

pilot light
and thermocouple

sacrificial
anode rod

temperature/
pressure
relief valve

discharge
pipe

baffle
(turbulator)

flue

drain
valve

burner

0 1 6 5

External (floating) tank water heater (oil-fired)

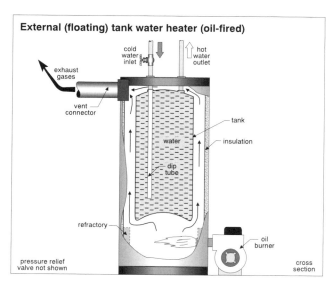

exhaust
gases

cold
water
inlet

hot
water
outlet

vent
connector

water

dip
tube

refractory

tank

insulation

oil
burner

pressure relief
valve not shown

cross
section

0 1 6 6

Dip tube

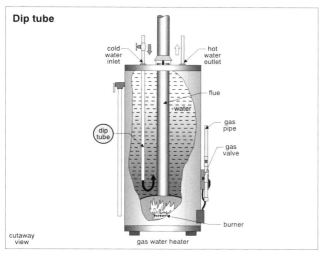

cold water inlet
hot water outlet
flue
water
gas pipe
dip tube
gas valve
burner
cutaway view
gas water heater

0 1 6 7

Recovery rate

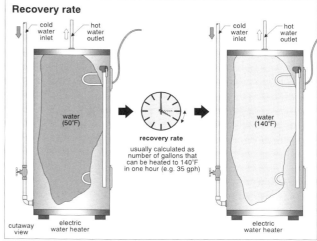

cold water inlet
hot water outlet
water (50°F)
cold water inlet
hot water outlet
water (140°F)

recovery rate
usually calculated as number of gallons that can be heated to 140°F in one hour (e.g. 35 gph)

cutaway view
electric water heater
electric water heater

0 1 6 8

Electric water heater

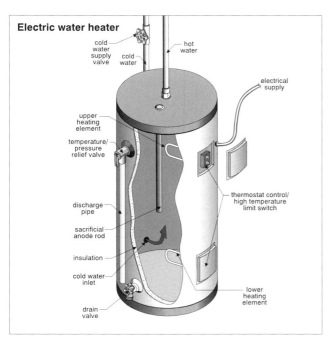

cold water supply valve
cold water
hot water
electrical supply
upper heating element
temperature/ pressure relief valve
discharge pipe
sacrificial anode rod
insulation
cold water inlet
drain valve
thermostat control/ high temperature limit switch
lower heating element

0 1 6 9

Electric water heater - element sequencing

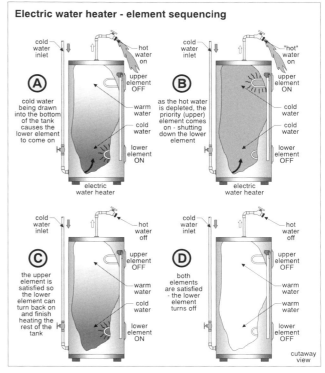

cold water inlet
hot water on
(A)
cold water being drawn into the bottom of the tank causes the lower element to come on
upper element OFF
warm water
cold water
lower element ON
electric water heater

cold water inlet
"hot" water on
(B)
as the hot water is depleted, the priority (upper) element comes on - shutting down the lower element
upper element ON
cold water
cold water
lower element OFF
electric water heater

cold water inlet
hot water off
(C)
the upper element is satisfied so the lower element can turn back on and finish heating the rest of the tank
upper element OFF
warm water
cold water
lower element ON

cold water inlet
hot water off
(D)
both elements are satisfied - the lower element turns off
upper element OFF
warm water
warm water
lower element OFF
cutaway view

0 1 7 0

Ice on regulator

ice build-up on regulators can block the vents and potentially allow excess gas pressure into the house

this is most likely to happen when the meter is below the drip line of the roof

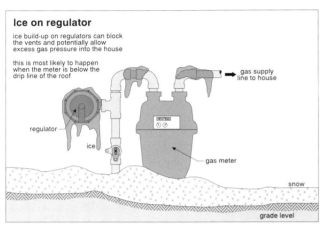

regulator

ice

gas supply line to house

gas meter

snow

grade level

0171

Poor meter locations

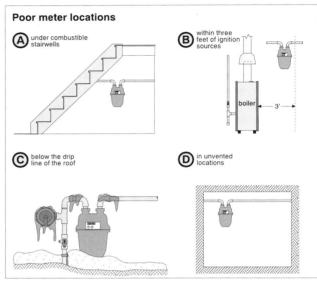

(A) under combustible stairwells

(B) within three feet of ignition sources

boiler

3'

(C) below the drip line of the roof

(D) in unvented locations

0172

Gas piping support
(steel pipe)

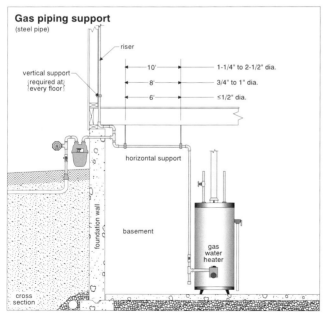

riser

vertical support (required at every floor)

10'	1-1/4" to 2-1/2" dia.
8'	3/4" to 1" dia.
6'	≤1/2" dia.

horizontal support

foundation wall

basement

gas water heater

cross section

0173

Drip leg

the drip leg (or dirt pocket) serves as a collection area for sediment to reduce the chance of clogged gas valves or burners

cold water supply valve

cold water

hot water

vent connector

draft hood

drip leg

gas pipe

gas valve

gas water heater

thermostat control

0174

Gas shut-off valves

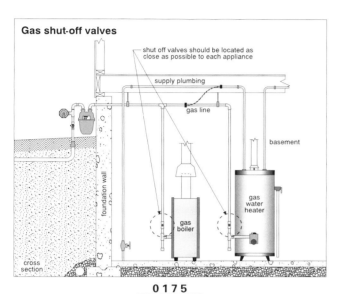

shut off valves should be located as close as possible to each appliance

supply plumbing

gas line

basement

foundation wall

gas boiler

gas water heater

cross section

0175

Teflon tape at connections

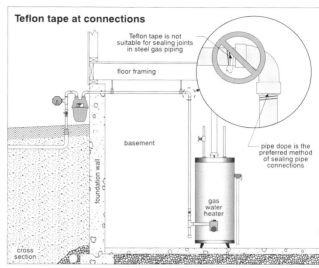

Teflon tape is not suitable for sealing joints in steel gas piping

floor framing

basement

foundation wall

gas water heater

pipe dope is the preferred method of sealing pipe connections

cross section

0176

Scorching

watch for evidence of scorching around the heat roll out shield, at the gas valve, at the gas tubing or on the outside of the tank

vent connector

hot water

draft hood

thermostat control

gas tubing

gas pipe

gas water heater

burner cover

heat roll out shield

0177

Vent connector length

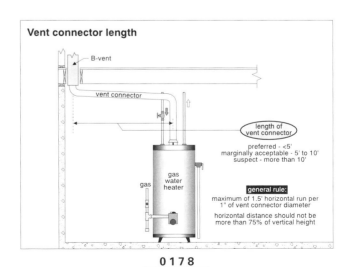

B-vent

vent connector

length of vent connector

preferred - <5'
marginally acceptable - 5' to 10'
suspect - more than 10'

general rule:

maximum of 1.5' horizontal run per 1" of vent connector diameter

horizontal distance should not be more than 75% of vertical height

gas

gas water heater

0178

Poor connections

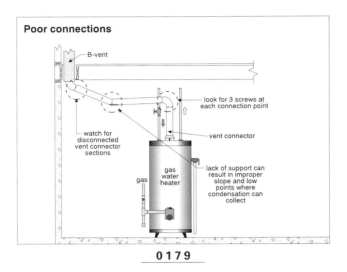

B-vent

look for 3 screws at each connection point

vent connector

watch for disconnected vent connector sections

gas

gas water heater

lack of support can result in improper slope and low points where condensation can collect

0179

Vent clearances

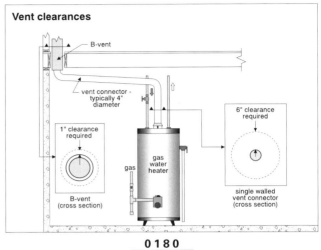

B-vent

vent connector - typically 4" diameter

1" clearance required

B-vent (cross section)

6" clearance required

gas

gas water heater

single walled vent connector (cross section)

0180

Size of vent connector

the vent connector diameter should match the size of the flue collar

if the vent connector is too large or too small, condensation or spillage could result

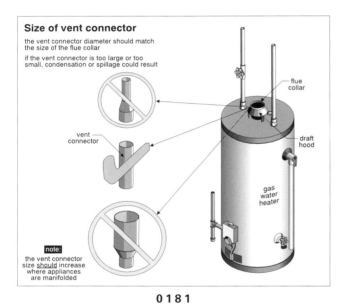

vent connector

flue collar

draft hood

gas water heater

note:
the vent connector size should increase where appliances are manifolded

0181

Chimney/vent connections

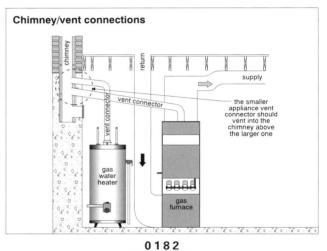

chimney

return

vent connector

supply

the smaller appliance vent connector should vent into the chimney above the larger one

gas water heater

gas furnace

0182

Vent connector extends too far into chimney

if the vent connector extends too far into the chimney, proper venting may be prevented

check for this in the chimney clean-out (with a mirror)

chimney

vent connector

gas

gas water heater

0183

Oil storage tanks - clearance from oil burner

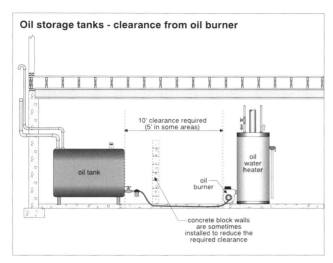

10' clearance required (5' in some areas)

oil tank

oil water heater

oil burner

concrete block walls are sometimes installed to reduce the required clearance

0 1 8 4

Oil storage tank leaks

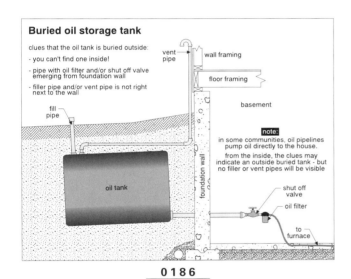

vent pipe

wall framing

floor framing

fill pipe

basement

oil gauge

watch these locations especially for oil leaks

foundation wall

oil tank

shut off valve

oil filter

0 1 8 5

Buried oil storage tank

clues that the oil tank is buried outside:

- you can't find one inside!

- pipe with oil filter and/or shut off valve emerging from foundation wall

- filler pipe and/or vent pipe is not right next to the wall

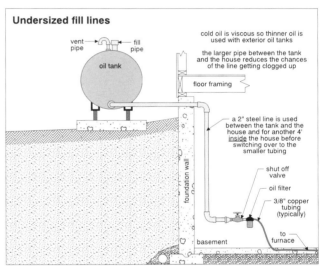

vent pipe

wall framing

floor framing

basement

note:

in some communities, oil pipelines pump oil directly to the house.

from the inside, the clues may indicate an outside buried tank - but no filler or vent pipes will be visible

fill pipe

foundation wall

oil tank

shut off valve

oil filter

to furnace

0 1 8 6

Undersized fill lines

vent pipe

fill pipe

oil tank

cold oil is viscous so thinner oil is used with exterior oil tanks

the larger pipe between the tank and the house reduces the chances of the line getting clogged up

floor framing

foundation wall

a 2" steel line is used between the tank and the house and for another 4' inside the house before switching over to the smaller tubing

shut off valve

oil filter

3/8" copper tubing (typically)

to furnace

basement

0 1 8 7

Primary controller

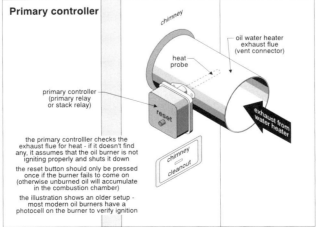

chimney

oil water heater exhaust flue (vent connector)

heat probe

primary controller (primary relay or stack relay)

reset

exhaust from water heater

chimney cleanout

the primary controlller checks the exhaust flue for heat - if it doesn't find any, it assumes that the oil burner is not igniting properly and shuts it down

the reset button should only be pressed once if the burner fails to come on (otherwise unburned oil will accumulate in the combustion chamber)

the illustration shows an older setup - most modern oil burners have a photocell on the burner to verify ignition

0 1 8 8

Refractory/fire pot

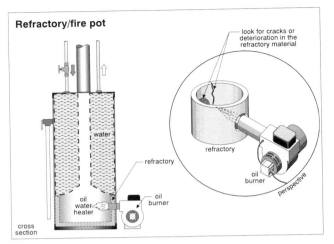

look for cracks or deterioration in the refractory material

water

refractory

refractory

oil burner

perspective

oil water heater

oil burner

cross section

0189

Barometric damper
(draft regulator)

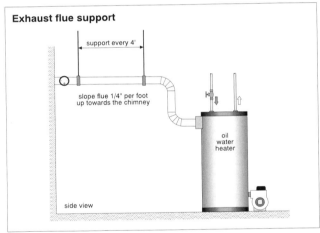

note:
not all oil fired water heaters will have a barometric damper

damper closed

damper open

draft air

counter-weight

view through flue

barometric damper (draft regulator)

exhaust flue

oil water heater

side view

0190

Exhaust flue support

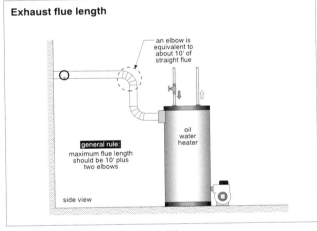

support every 4'

slope flue 1/4" per foot up towards the chimney

oil water heater

side view

0191

Exhaust flue length

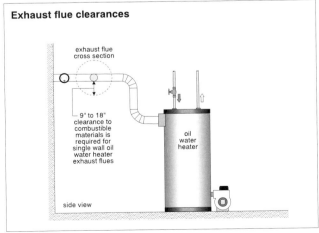

an elbow is equivalent to about 10' of straight flue

general rule:
maximum flue length should be 10' plus two elbows

oil water heater

side view

0192

Exhaust flue clearances

exhaust flue cross section

9" to 18" clearance to combustible materials is required for single wall oil water heater exhaust flues

oil water heater

side view

0193

Relative recovery rates

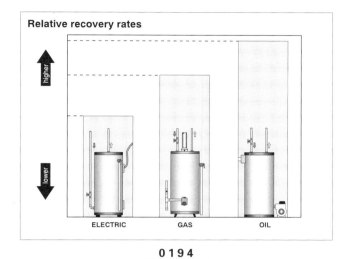

higher

lower

ELECTRIC   GAS   OIL

0194

BTUs per hour compared to Kilowatts

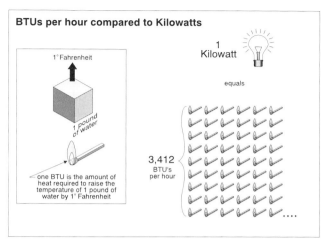

1° Fahrenheit

1 pound of water

one BTU is the amount of heat required to raise the temperature of 1 pound of water by 1° Fahrenheit

1 Kilowatt

equals

3,412 BTU's per hour

. . . .

0195

Gas burner cover and roll-out shield

check to see that the burner cover and heat roll out shield are both present and in good condition

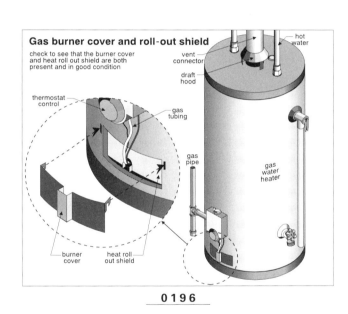

hot water

vent connector

draft hood

thermostat control

gas tubing

gas pipe

gas water heater

burner cover

heat roll out shield

0196

Working space around water heaters

adequate space (24") should be provided around the water heater to allow for servicing and replacement

doors into these areas should be wide enough to allow for removal of the tank

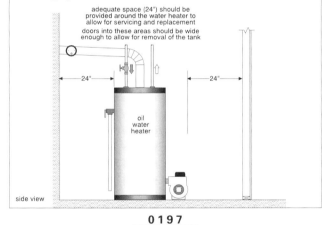

24"   24"

oil water heater

side view

0197

Water heaters in garages

gas or fired water heaters that are located in a garage must be at least 18" above the floor so that gasoline vapors are not ignited by the pilot or burner

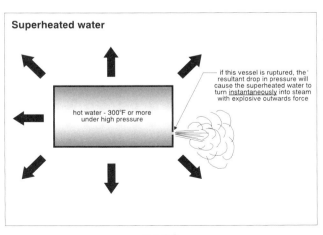

oil (or gas) water heater

18"

side view

0198

Superheated water

if this vessel is ruptured, the resultant drop in pressure will cause the superheated water to turn instantaneously into steam with explosive outwards force

hot water - 300°F or more under high pressure

0199

Temperature/pressure relief valve

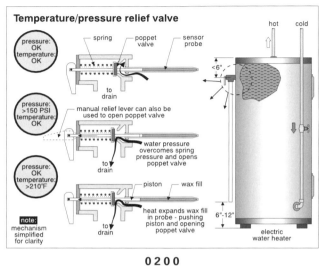

pressure: OK temperature: OK

spring — poppet valve — sensor probe

hot   cold

to drain

pressure: >150 PSI temperature: OK

manual relief lever can also be used to open poppet valve

water pressure overcomes spring pressure and opens poppet valve

to drain

pressure: OK temperature: >210°F

piston   wax fill

heat expands wax fill in probe - pushing piston and opening poppet valve

to drain

note: mechanism simplified for clarity

6"-12"

electric water heater

0200

Gas shut-off on high temperature

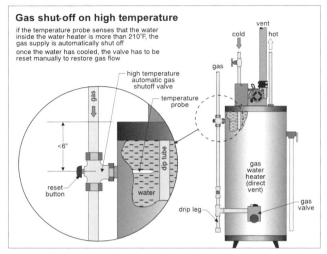

if the temperature probe senses that the water inside the water heater is more than 210°F, the gas supply is automatically shut off

once the water has cooled, the valve has to be reset manually to restore gas flow

vent
cold   hot
gas

high temperature automatic gas shutoff valve

temperature probe

gas

dip tube

water

reset button

gas water heater (direct vent)

drip leg

gas valve

0201

Baffle collapsed or missing

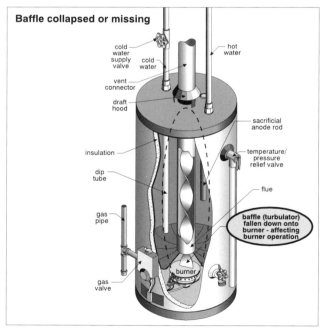

cold water supply valve
cold water
hot water

vent connector

draft hood

sacrificial anode rod

insulation

temperature/ pressure relief valve

dip tube

flue

gas pipe

baffle (turbulator) fallen down onto burner - affecting burner operation

burner

gas valve

0202

Isolating and drain valves

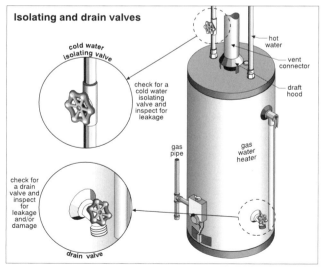

cold water isolating valve

hot water

vent connector

check for a cold water isolating valve and inspect for leakage

draft hood

gas pipe

gas water heater

check for a drain valve and inspect for leakage and/or damage

drain valve

0203

Multiple water heaters are installed in parallel

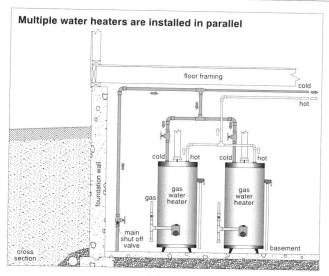

0204

Circulating hot water system

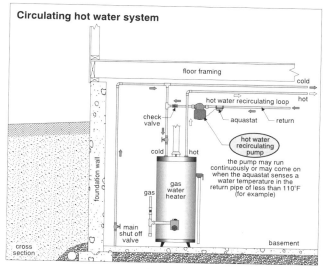

0205

Power vented water heater

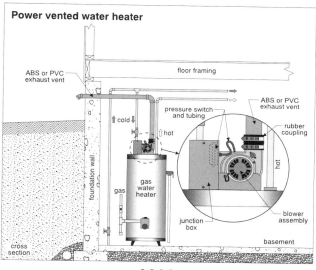

0206

Improper side wall vent locations

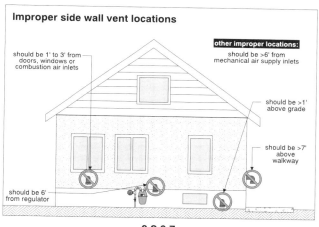

0207

High-efficiency gas water heater

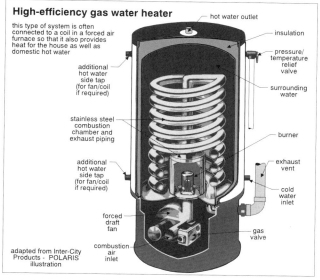

0208

Combination furnace/water heater system

0209

Tempering valve required

in order to maximize the capacity of the combination system, the water heater temperature is sometimes turned way up

a **tempering valve** adds a little cold water to the very hot water coming out of the water heater so that it is cool enough for domestic use

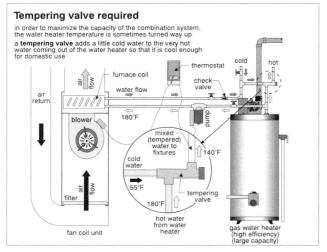

0 2 1 0

Boilers heat domestic water

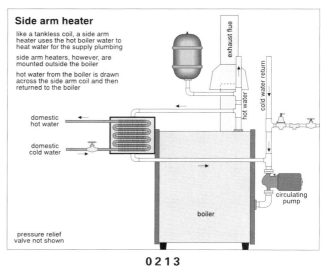

0 2 1 1

Tankless coil

a tankless coil uses the hot boiler water to heat water for the supply plumbing

it is a slide-in option for some boilers

to transfer heat from the <u>hottest</u> water, it is located near the top of the boiler

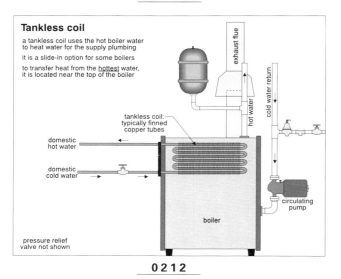

0 2 1 2

Side arm heater

like a tankless coil, a side arm heater uses the hot boiler water to heat water for the supply plumbing

side arm heaters, however, are mounted outside the boiler

hot water from the boiler is drawn across the side arm coil and then returned to the boiler

0 2 1 3

Tempering valve with tankless coil

a tempering valve mixes some of the incoming domestic cold water with the "too hot" water coming out of the tankless coil to bring it down to a temperature suitable for domestic use

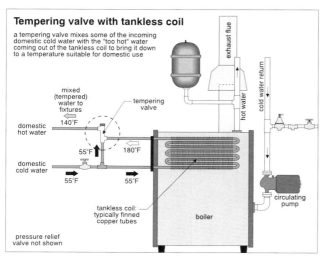

0 2 1 4

Instantaneous water heater

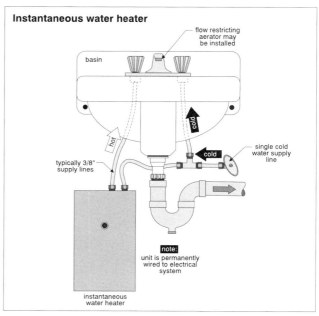

0 2 1 5

Too little or too much slope isn't good

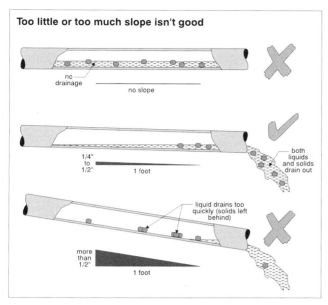

no drainage
no slope

1/4" to 1/2"
1 foot

both liquids and solids drain out

liquid drains too quickly (solids left behind)

more than 1/2"
1 foot

0216

Trap terminology

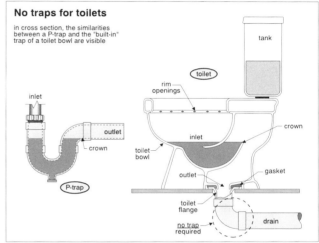

inlet
tail piece (or fixture outlet pipe)
trap adapter
outlet
trap arm (or fixture drain)
trap dip (upper dip)
trap seal depth
trap weir (crown weir)
trap seal depth must be at least 1-1/2" to 2" (and not more than 4" according to some authorities)
P-trap (with cleanout)
cleanout
lower dip

cross section

0217

Trap cleanout required

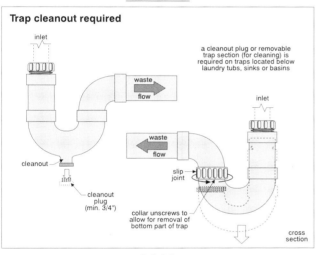

inlet

a cleanout plug or removable trap section (for cleaning) is required on traps located below laundry tubs, sinks or basins

waste flow

inlet

cleanout

waste flow

cleanout plug (min. 3/4")

slip joint

collar unscrews to allow for removal of bottom part of trap

cross section

0218

No traps for toilets

in cross section, the similarities between a P-trap and the "built-in" trap of a toilet bowl are visible

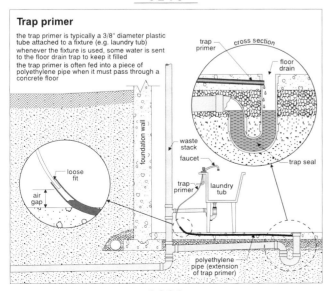

inlet
outlet
crown
P-trap

tank
toilet
rim openings
crown
inlet
toilet bowl
outlet
gasket
toilet flange
no trap required
drain

0219

Trap primer

the trap primer is typically a 3/8" diameter plastic tube attached to a fixture (e.g. laundry tub)
whenever the fixture is used, some water is sent to the floor drain trap to keep it filled
the trap primer is often fed into a piece of polyethylene pipe when it must pass through a concrete floor

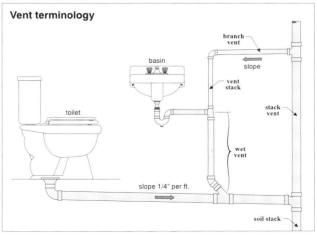

cross section
trap primer
floor drain
foundation wall
waste stack
faucet
trap primer
laundry tub
trap seal
loose fit
air gap
polyethylene pipe (extension of trap primer)

0220

Vent terminology

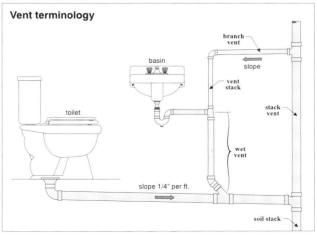

branch vent
slope
basin
vent stack
stack vent
toilet
wet vent
slope 1/4" per ft.
soil stack

0221

Soil stack versus waste stack

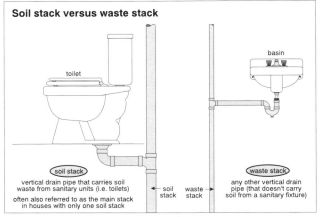

basin

toilet

soil stack

waste stack

soil stack

waste stack

vertical drain pipe that carries soil waste from sanitary units (i.e. toilets)

often also referred to as the main stack in houses with only one soil stack

any other vertical drain pipe (that doesn't carry soil from a sanitary fixture)

0222

Stack vent

the stack vent is an extension of the waste stack that runs up through the roof to the exterior - for venting of exhaust gases and to maintain atmospheric pressure in the waste system

vent

stack flashing

ceiling joist

roof rafter

stack vent

basin

wall framing

waste stack

0223

Proper vent location relative to trap

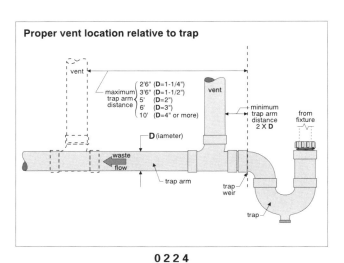

vent

maximum trap arm distance
- 2'6" (**D**=1-1/4")
- 3'6" (**D**=1-1/2")
- 5'   (**D**=2")
- 6'   (**D**=3")
- 10'   (**D**=4" or more)

vent

minimum trap arm distance 2 X **D**

from fixture

D (iameter)

waste flow

trap arm

trap weir

trap

0224

Direct venting

if a fixture drain is located within 5 feet of a waste stack, a standard vent is often not required as the stack itself serves as the vent

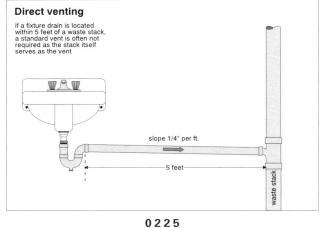

slope 1/4" per ft.

5 feet

waste stack

0225

Wet venting

toilet wet vented by basin

dry vent

slope

basin

stack vent

toilet

wet vent (one size larger than normal)

slope 1/4" per ft.

soil stack

0226

Connecting cast iron waste pipe - hubless

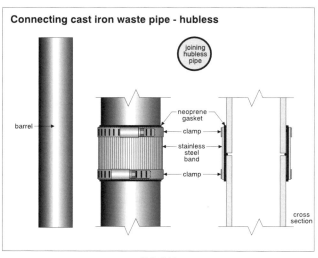

joining hubless pipe

barrel

neoprene gasket

clamp

stainless steel band

clamp

cross section

0227

Cast iron waste pipe - bell and spigot
(sometimes called hub and spigot)

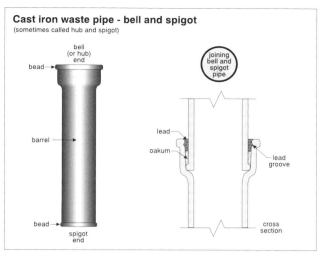

bell (or hub) end

bead

barrel

bead

spigot end

joining bell and spigot pipe

lead

oakum

lead groove

cross section

0228

Cross connections

cross connections can occur in fixtures where the faucet is below the high water level - an air gap is required to prevent contaminated water from being siphoned back into the potable water supply under negative supply pressure conditions

bidet

faucet

solution
install vacuum breaker

older style bathtub with low faucet

faucet

solution
replace fixture

faucet

flood level rim

air gap

solution

laundry tub

replace faucet

0229

Standpipe

clothes washer

indirect connection

clothes washer drain hose

standpipe

18" min. to 30" max. (some areas)

trap (may not be permitted below floor)

6" min. to 18" max.

waste stack

0230

Condensate discharge locations

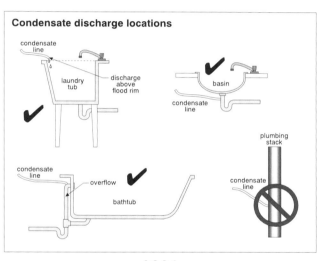

condensate line

laundry tub

discharge above flood rim

basin

condensate line

condensate line

overflow

bathtub

plumbing stack

condensate line

0231

Connecting vertical drain pipes to horizontal drain pipes

vertical drain from fixture above

vertical drain from fixture above

horizontal waste pipe

slope: 1/4" per ft.

horizontal waste pipe

slope: 1/4" per ft.

TY fitting is **not** acceptable

a **Y** fitting is correct for this connection

note:
a TY fitting can be used to connect a **vent** pipe to horizontal waste pipe

0232

Building trap - older installation

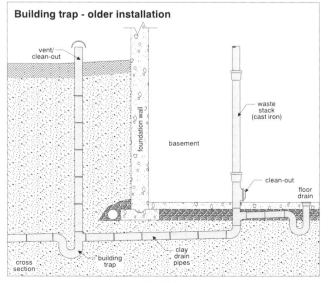

vent/ clean-out

foundation wall

basement

waste stack (cast iron)

clean-out

floor drain

cross section

building trap

clay drain pipes

0233

Building trap - new installation

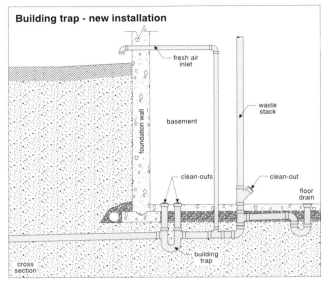

fresh air inlet

foundation wall

basement

waste stack

clean-outs

clean-out

floor drain

building trap

cross section

0234

Illegal traps

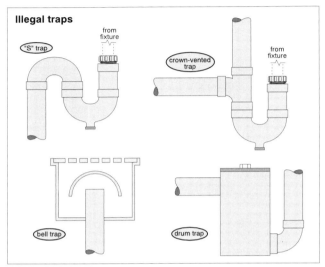

from fixture

"S" trap

crown-vented trap

from fixture

bell trap

drum trap

0235

Double trapping doesn't work

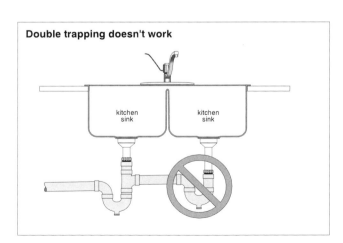

kitchen sink

kitchen sink

0236

Tail piece too long

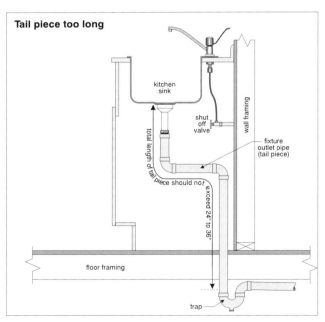

kitchen sink

wall framing

shut off valve

fixture outlet pipe (tail piece)

total length of tail piece should not exceed 24" to 36"

floor framing

trap

0237

Venting an island sink (below-floor trap permitted)

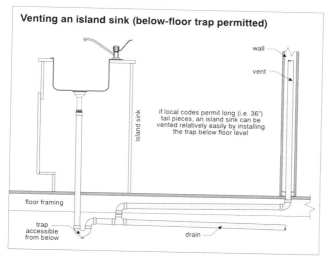

wall

vent

island sink

if local codes permit long (i.e. 36")
tail pieces, an island sink can be
vented relatively easily by installing
the trap below floor level

floor framing

trap
accessible
from below

drain

0238

Loop vent or circuit vent for an island sink

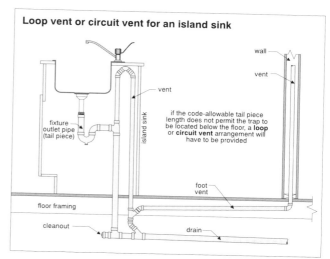

wall

vent

vent

fixture
outlet pipe
(tail piece)

island sink

if the code-allowable tail piece
length does not permit the trap to
be located below the floor, a **loop**
or **circuit vent** arrangement will
have to be provided

foot
vent

floor framing

cleanout

drain

0239

Backwater valve

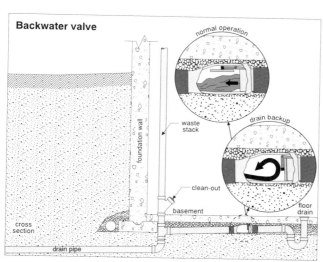

normal operation

drain backup

foundation wall

waste
stack

clean-out

basement

floor
drain

cross
section

drain pipe

0240

Downspout connection upstream of trap

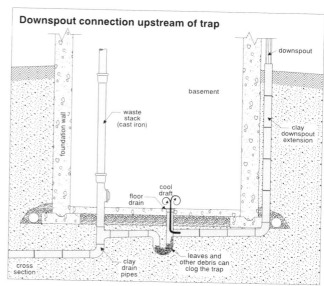

downspout

basement

foundation wall

waste
stack
(cast iron)

clay
downspout
extension

cool
draft

floor
drain

clay
drain
pipes

leaves and
other debris can
clog the trap

cross
section

0241

Flashing problems

flashing leaks can look like plumbing
leaks because the water tends to cling to
the outside of the stack as it drips down
- until it hits an obstruction

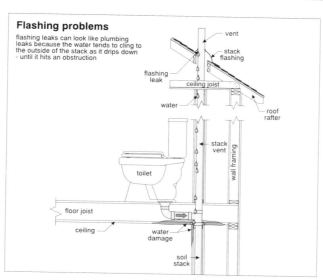

vent

stack
flashing

flashing
leak

ceiling joist

water

roof
rafter

stack
vent

toilet

wall framing

floor joist

ceiling

water
damage

soil
stack

0242

Horizontal vent offset

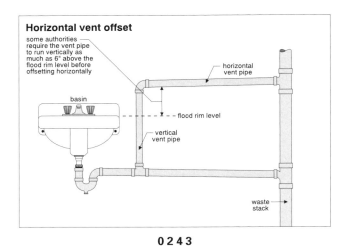

some authorities require the vent pipe to run vertically as much as 6" above the flood rim level before offsetting horizontally

horizontal vent pipe

basin

flood rim level

vertical vent pipe

waste stack

0 2 4 3

Vent on the wrong side of the trap

vent

sewer gas

slope

basin

stack

slope 1/4" per ft.

0 2 4 4

Venting an island sink (if below-floor trap not permitted)

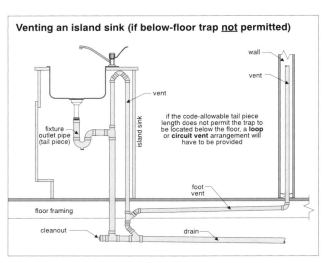

wall

vent

vent

if the code-allowable tail piece length does not permit the trap to be located below the floor, a **loop** or **circuit vent** arrangement will have to be provided

fixture outlet pipe (tail piece)

island sink

foot vent

floor framing

cleanout

drain

0 2 4 5

Another alternate island venting arrangement

not permitted in all areas

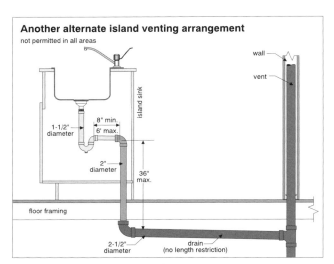

wall

vent

island sink

1-1/2" diameter

8" min.
6' max.

2" diameter

36" max.

floor framing

2-1/2" diameter

drain (no length restriction)

0 2 4 6

Vent too tall

if the plumbing vent extends too far above the roof, frost closure can become a problem

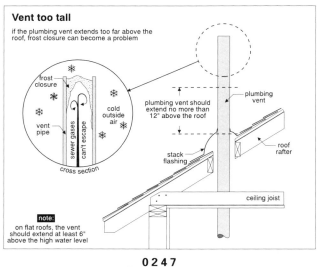

frost closure

cold outside air

vent pipe

sewer gases can't escape

cross section

plumbing vent should extend no more than 12" above the roof

plumbing vent

stack flashing

roof rafter

ceiling joist

note:
on flat roofs, the vent should extend at least 6" above the high water level

0 2 4 7

Plumbing vent clearances

3 to 6 feet from property line

3 feet above operable doors or windows

7 feet above decks

10 to 12 feet from doors and windows at the same elevation

7 to 10 feet above grade

0 2 4 8

Automatic air vent or air admittance valve

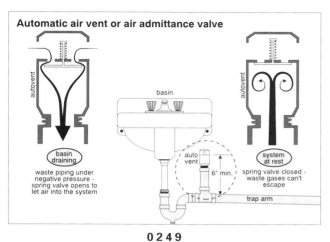

autovent

basin

autovent

basin draining

auto vent

6" min.

system at rest

waste piping under negative pressure - spring valve opens to let air into the system

spring valve closed - waste gases can't escape

trap arm

0249

Sewage ejector pump

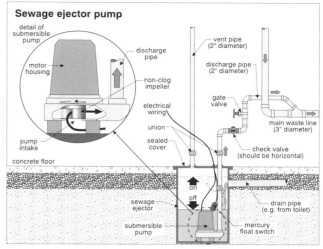

detail of submersible pump

motor housing

discharge pipe

non-clog impeller

electrical wiring

union

sealed cover

pump intake

concrete floor

sewage ejector

submersible pump

mercury float switch

on

off

vent pipe (2" diameter)

discharge pipe (2" diameter)

gate valve

main waste line (3" diameter)

check valve (should be horizontal)

drain pipe (e.g. from toilet)

0250

Sump pump
pedestal type

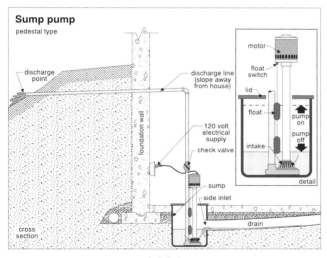

discharge point

discharge line (slope away from house)

foundation wall

120 volt electrical supply

check valve

sump

side inlet

drain

cross section

motor

float switch

lid

float

pump on

pump off

intake

detail

0251

Laundry tub pump

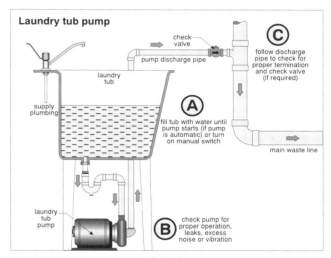

check valve

pump discharge pipe

laundry tub

supply plumbing

laundry tub pump

C follow discharge pipe to check for proper termination and check valve (if required)

A fill tub with water until pump starts (if pump is automatic) or turn on manual switch

main waste line

B check pump for proper operation, leaks, excess noise or vibration

0252

Septic tank
(two compartment)

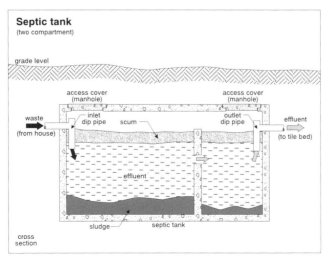

grade level

access cover (manhole)

access cover (manhole)

waste (from house)

inlet dip pipe

scum

outlet dip pipe

effluent (to tile bed)

effluent

sludge

septic tank

cross section

0253

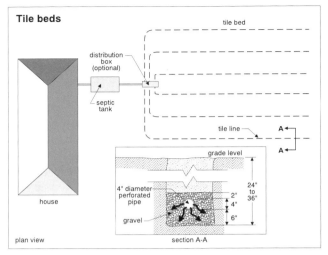

Tile beds

distribution box (optional)

septic tank

house

tile bed

tile line

A

grade level

4" diameter perforated pipe

gravel

24" to 36"

2"

4"

6"

plan view

section A-A

0254

Tile bed location

property line

5' min.

10' min.

tile bed

6' to 10'

tile line - 100' max.

10' min.

50' min.

lake

septic tank

house

15' min.

50' minimum if well has a casing
100' minimum if well has *no* casing

well

plan view

0255

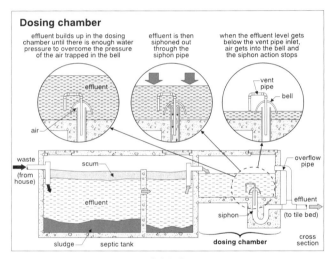

Dosing chamber

effluent builds up in the dosing chamber until there is enough water pressure to overcome the pressure of the air trapped in the bell

effluent is then siphoned out through the siphon pipe

when the effluent level gets below the vent pipe inlet, air gets into the bell and the siphon action stops

effluent

air

vent pipe

bell

waste (from house)

scum

effluent

sludge

septic tank

siphon

overflow pipe

effluent (to tile bed)

dosing chamber

cross section

0256

Seepage pit

some seepage pits have masonry or stone walls but function similarly

concrete seepage pit

2 feet of gravel all around outside of seepage pit

grade level

top soil

outlet pipe from septic tank

effluent

gravel fill

0257

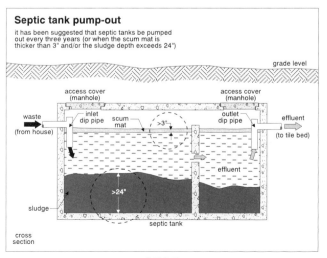

Septic tank pump-out

it has been suggested that septic tanks be pumped out every three years (or when the scum mat is thicker than 3" and/or the sludge depth exceeds 24")

grade level

access cover (manhole)

access cover (manhole)

waste (from house)

inlet dip pipe

scum mat

>3"

outlet dip pipe

effluent (to tile bed)

effluent

sludge

>24"

septic tank

cross section

0258

Hot/cold conventions

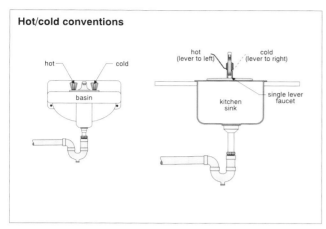

hot — cold
basin

hot (lever to left) — cold (lever to right)
kitchen sink
single lever faucet

0259

Rusting overflow

rust starts to develop where the overflow is spot welded to the basin

the rust can spread and ultimately eat through the basin (or overflow) causing leakage

the spot weld areas are also where (and why) enamel on the inside of the basin tends to chip

on newer enameled steel basins, the overflow is <u>siliconed</u> onto the basin

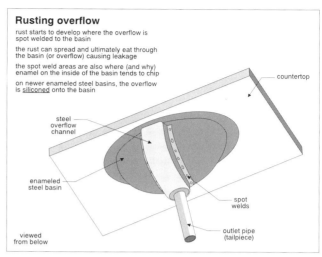

steel overflow channel
countertop
enameled steel basin
spot welds
outlet pipe (tailpiece)
viewed from below

0260

Compression faucet

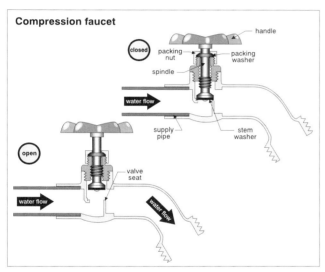

handle
closed
packing nut
packing washer
spindle
water flow
supply pipe
stem washer
open
valve seat
water flow
water flow

0261

Shut-off valves for outside faucets

piping/faucets downstream of the shutoff valve should be sloped downwards for drainage

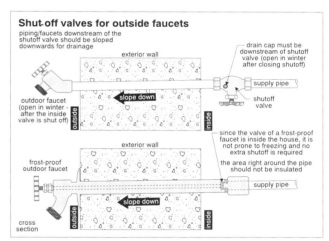

exterior wall
drain cap must be downstream of shutoff valve (open in winter after closing shutoff)
slope down
supply pipe
outdoor faucet (open in winter - after the inside valve is shut off)
outside
inside
shutoff valve

since the valve of a frost-proof faucet is inside the house, it is not prone to freezing and no extra shutoff is required

the area right around the pipe should not be insulated

exterior wall
frost-proof outdoor faucet
slope down
supply pipe
outside
inside
cross section

0262

Toilet flushing actions

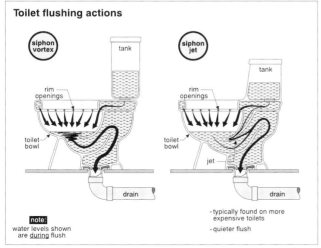

siphon vortex
tank
rim openings
toilet bowl
drain

siphon jet
tank
rim openings
toilet bowl
jet
drain

- typically found on more expensive toilets
- quieter flush

note:
water levels shown are <u>during</u> flush

0263

Supply tubes

supply tubes are used to connect water pipes to faucets, toilets and other fixtures

they can be plastic (semi-rigid), chromed copper (semi-rigid), braided steel (flexible) or vinyl mesh (flexible)

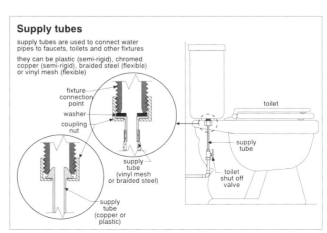

fixture connection point
washer
coupling nut
supply tube (vinyl mesh or braided steel)
supply tube (copper or plastic)
toilet
supply tube
toilet shut off valve

0264

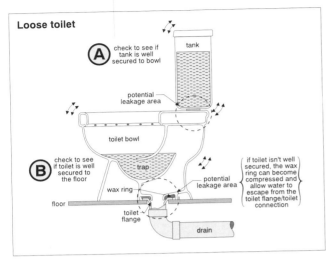

Loose toilet

Ⓐ check to see if tank is well secured to bowl

tank

potential leakage area

toilet bowl

trap

Ⓑ check to see if toilet is well secured to the floor

wax ring

floor

toilet flange

drain

potential leakage area

if toilet isn't well secured, the wax ring can become compressed and allow water to escape from the toilet flange/toilet connection

0265

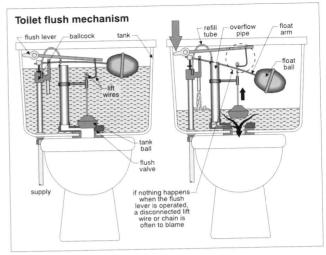

Toilet flush mechanism

flush lever — ballcock — tank

refill tube — overflow pipe — float arm

float ball

lift wires

tank ball

flush valve

supply

if nothing happens when the flush lever is operated, a disconnected lift wire or chain is often to blame

0266

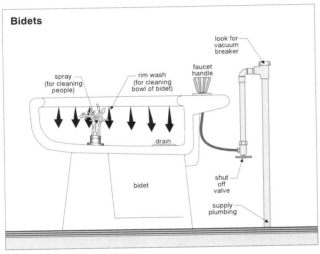

Bidets

look for vacuum breaker

spray (for cleaning people)

rim wash (for cleaning bowl of bidet)

faucet handle

drain

bidet

shut off valve

supply plumbing

0267

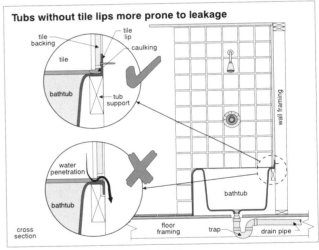

Tubs without tile lips more prone to leakage

tile backing

tile lip

caulking

tile

bathtub

tub support

wall framing

water penetration

bathtub

bathtub

cross section

floor framing

trap

drain pipe

0268

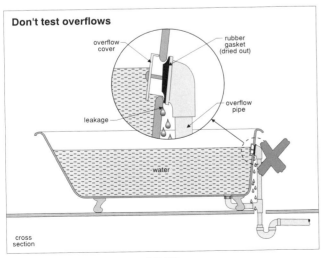

Don't test overflows

overflow cover

rubber gasket (dried out)

leakage

overflow pipe

water

cross section

0269

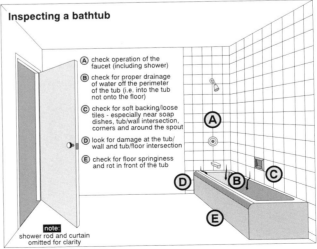

Inspecting a bathtub

Ⓐ check operation of the faucet (including shower)

Ⓑ check for proper drainage of water off the perimeter of the tub (i.e. into the tub not onto the floor)

Ⓒ check for soft backing/loose tiles - especially near soap dishes, tub/wall intersection, corners and around the spout

Ⓓ look for damage at the tub/wall and tub/floor intersection

Ⓔ check for floor springiness and rot in front of the tub

note: shower rod and curtain omitted for clarity

0270

Bathtubs on exterior walls

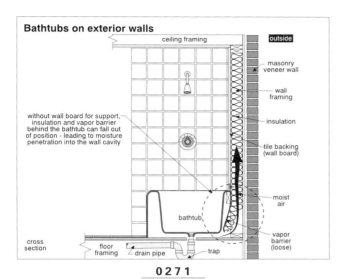

ceiling framing

outside

masonry veneer wall

wall framing

insulation

without wall board for support, insulation and vapor barrier behind the bathtub can fall out of position - leading to moisture penetration into the wall cavity

tile backing (wall board)

moist air

bathtub

vapor barrier (loose)

cross section

floor framing

drain pipe   trap

0271

Bathtub windows aren't a good idea

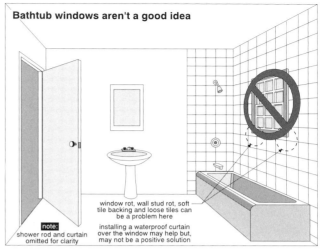

window rot, wall stud rot, soft tile backing and loose tiles can be a problem here

note: shower rod and curtain omitted for clarity

installing a waterproof curtain over the window may help but, may not be a positive solution

0272

Bathroom outlets and switches

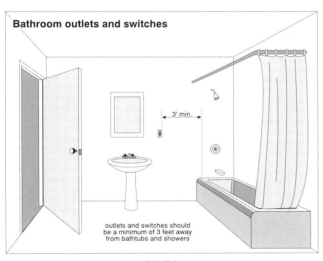

3' min.

outlets and switches should be a minimum of 3 feet away from bathtubs and showers

0273

One-piece shower stalls

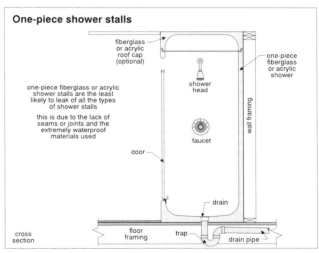

fiberglass or acrylic roof cap (optional)

one-piece fiberglass or acrylic shower

one-piece fiberglass or acrylic shower stalls are the least likely to leak of all the types of shower stalls

shower head

wall framing

this is due to the lack of seams or joints and the extremely waterproof materials used

door

faucet

drain

cross section

floor framing   trap

drain pipe

0274

Acrylic base

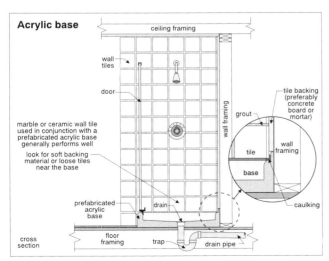

ceiling framing

wall tiles

door

tile backing (preferably concrete board or mortar)

grout

marble or ceramic wall tile used in conjunction with a prefabricated acrylic base generally performs well

tile

wall framing

look for soft backing material or loose tiles near the base

base

prefabricated acrylic base

drain

caulking

cross section

floor framing   trap

drain pipe

0275

Metal showers are bad

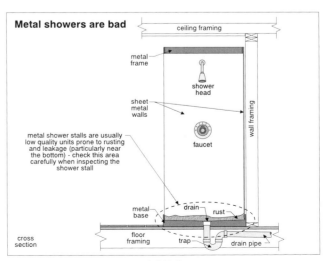

ceiling framing

metal frame

shower head

sheet metal walls

wall framing

metal shower stalls are usually low quality units prone to rusting and leakage (particularly near the bottom) - check this area carefully when inspecting the shower stall

faucet

metal base   drain   rust

cross section

floor framing   trap

drain pipe

0276

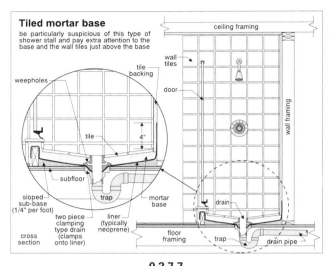

Tiled mortar base

be particularly suspicious of this type of shower stall and pay extra attention to the base and the wall tiles just above the base

ceiling framing

wall tiles

tile backing

weepholes

door

tile

4"

wall framing

subfloor

sloped sub-base (1/4" per foot)

trap

mortar base

two piece clamping type drain (clamps onto liner)

liner (typically neoprene)

drain

cross section

floor framing

trap

drain pipe

0277

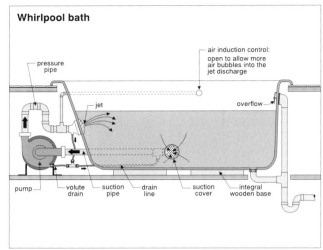

Whirlpool bath

pressure pipe

air induction control: open to allow more air bubbles into the jet discharge

jet

overflow

pump

volute drain

suction pipe

drain line

suction cover

integral wooden base

0278

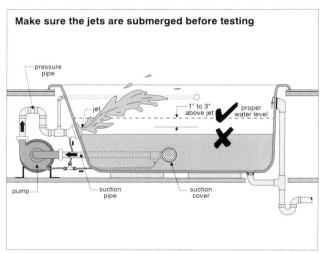

Make sure the jets are submerged before testing

pressure pipe

jet

1" to 3" above jet

proper water level

pump

suction pipe

suction cover

0279

PART 3

INSULATION

THE BASICS

AIR/VAPOR BARRIER (VAPOR RETARDERS)

VENTING ROOFS

VENTING LIVING SPACES

ATTICS

FLAT AND CATHEDRAL ROOFS

British Thermal Unit (BTUs)

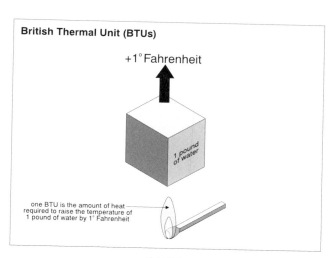

+1°Fahrenheit

1 pound of water

one BTU is the amount of heat required to raise the temperature of 1 pound of water by 1° Fahrenheit

0280

Latent heat of vaporization

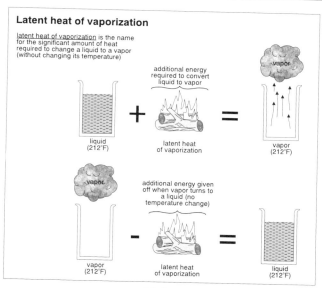

latent heat of vaporization is the name for the significant amount of heat required to change a liquid to a vapor (without changing its temperature)

liquid (212°F) + latent heat of vaporization = vapor (212°F)

additional energy required to convert liquid to vapor

vapor (212°F) − latent heat of vaporization = liquid (212°F)

additional energy given off when vapor turns to a liquid (no temperature change)

0281

Mechanisms of heat transfer

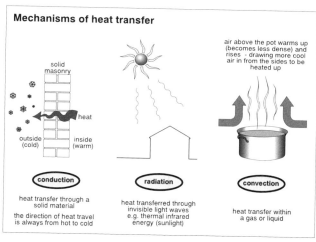

solid masonry

heat

outside (cold)   inside (warm)

conduction

heat transfer through a solid material

the direction of heat travel is always from hot to cold

radiation

heat transferred through invisible light waves e.g. thermal infrared energy (sunlight)

air above the pot warms up (becomes less dense) and rises - drawing more cool air in from the sides to be heated up

convection

heat transfer within a gas or liquid

0282

People and homes release heat

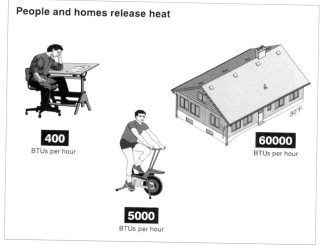

400
BTUs per hour

5000
BTUs per hour

30°F

60000
BTUs per hour

0283

Thermal conductivity

the thermal conductivity (k) of a homogeneous material is equal to the number of BTUs that will pass through one square foot of the material (that is 1" thick) over the course of 1 hour (with a 1 °F temperature difference across the material)

in this case, the thermal conductivity (k) of the concrete slab is 12

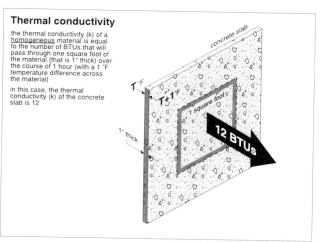

concrete slab

T °F
T+1°F
1 square foot
1" thick
12 BTUs

0284

Conductance

the conductance (C) is equal to the number of BTUs that will pass through one square foot of the material over the course of 1 hour (with a 1 °F temperature difference across the material)

in this case, the conductance (C) of the concrete block wall is 0.9

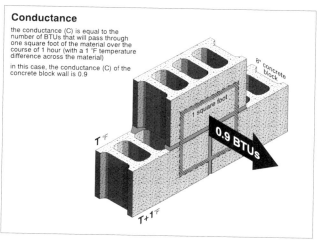

8" concrete block

T °F
1 square foot
0.9 BTUs
T+1°F

0285

Two comfort strategies

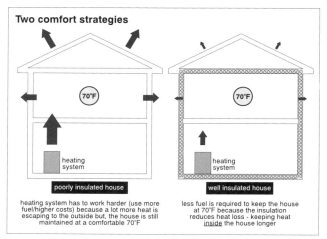

70°F

heating system

poorly insulated house

heating system has to work harder (use more fuel/higher costs) because a lot more heat is escaping to the outside but, the house is still maintained at a comfortable 70°F

70°F

heating system

well insulated house

less fuel is required to keep the house at 70°F because the insulation reduces heat loss - keeping heat <u>inside</u> the house longer

0286

Convective loop

air

convective loop

cool

cool

warm air

cool air

insulation holds the air in pockets so small that there can be no movement of air (preventing the formation of convective loops)

insulation

cool

cool

warm air

0287

R-value per inch

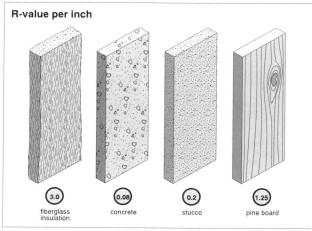

(3.0) fiberglass insulation

(0.08) concrete

(0.2) stucco

(1.25) pine board

0288

Recommended insulation levels
(northern North America)

attic

R-25 to R-40

R-20 to R-30

floors over unheated spaces

R-12 to R-20 walls

0289

Balanced air changes

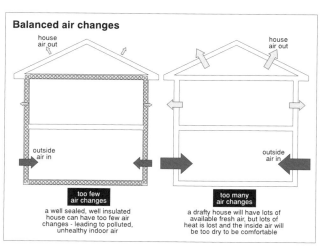

house air out

outside air in

too few air changes

a well sealed, well insulated house can have too few air changes - leading to polluted, unhealthy indoor air

house air out

outside air in

too many air changes

a drafty house will have lots of available fresh air, but lots of heat is lost and the inside air will be too dry to be comfortable

0290

Absolute and relative humidity

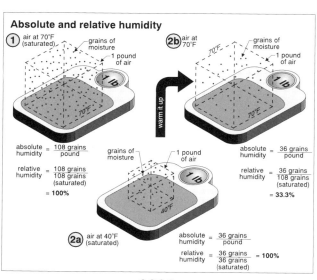

(1) air at 70°F (saturated)

grains of moisture

1 pound of air

1 lb

70°F

$\text{absolute humidity} = \dfrac{108 \text{ grains}}{\text{pound}}$

$\text{relative humidity} = \dfrac{108 \text{ grains}}{108 \text{ grains}} \text{ (saturated)}$

$= 100\%$

(2b) air at 70°F

grains of moisture

1 pound of air

1 lb

70°F

warm it up

$\text{absolute humidity} = \dfrac{36 \text{ grains}}{\text{pound}}$

$\text{relative humidity} = \dfrac{36 \text{ grains}}{108 \text{ grains}} \text{ (saturated)}$

$= 33.3\%$

grains of moisture

1 pound of air

1 lb

40°F

(2a) air at 40°F (saturated)

$\text{absolute humidity} = \dfrac{36 \text{ grains}}{\text{pound}}$

$\text{relative humidity} = \dfrac{36 \text{ grains}}{36 \text{ grains}} \text{ (saturated)} = 100\%$

0291

How moisture moves

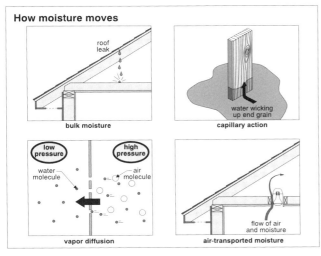

bulk moisture

roof leak

capillary action

water wicking up end grain

vapor diffusion

low pressure / high pressure
water molecule / air molecule

air-transported moisture

flow of air and moisture

0292

Stack effect

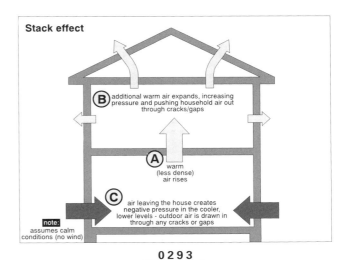

B additional warm air expands, increasing pressure and pushing household air out through cracks/gaps

A warm (less dense) air rises

C air leaving the house creates negative pressure in the cooler, lower levels - outdoor air is drawn in through any cracks or gaps

note: assumes calm conditions (no wind)

0293

Neutral pressure plane

outside pressure

exfiltrating air

higher pressure

neutral pressure plane

exfiltrating air

lower pressure

infiltrating air

infiltrating air

note: assumes calm conditions (no wind)

0294

Dew point

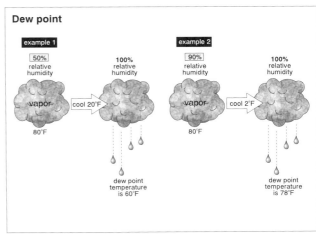

example 1

50% relative humidity

vapor

80°F

cool 20°F

100% relative humidity

dew point temperature is 60°F

example 2

90% relative humidity

vapor

80°F

cool 2°F

100% relative humidity

dew point temperature is 78°F

0295

Vapor diffusion

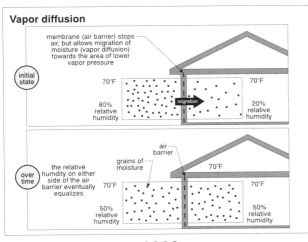

membrane (air barrier) stops air, but allows migration of moisture (vapor diffusion) towards the area of lower vapor pressure

initial state

70°F   70°F
80% relative humidity   20% relative humidity
migration

over time

the relative humidty on either side of the air barrier eventually equalizes

air barrier

grains of moisture

70°F   70°F
50% relative humidity   50% relative humidity

0296

Drafty is good (for houses)

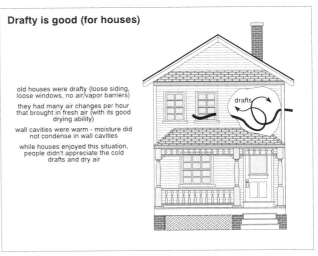

old houses were drafty (loose siding, loose windows, no air/vapor barriers)

they had many air changes per hour that brought in fresh air (with its good drying ability)

wall cavities were warm - moisture did not condense in wall cavities

while houses enjoyed this situation, people didn't appreciate the cold drafts and dry air

drafts

0297

Adding new siding can trap moisture

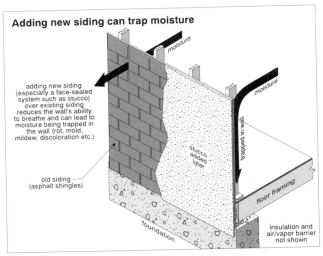

adding new siding (especially a face-sealed system such as stucco) over existing siding reduces the wall's ability to breathe and can lead to moisture being trapped in the wall (rot, mold, mildew, discoloration etc.)

moisture

moisture

trapped in wall

old siding (asphalt shingles)

stucco added later

floor framing

foundation

insulation and air/vapor barrier not shown

0 2 9 8

Moisture control in hot climates

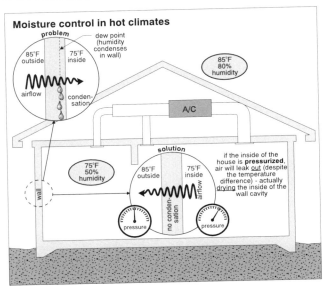

problem

dew point (humidity condensity in wall)

85°F outside

75°F inside

airflow

conden-sation

85°F 80% humidity

A/C

solution

75°F 50% humidity

wall

85°F outside

75°F inside

airflow

no conden-sation

pressure

pressure

if the inside of the house is **pressurized**, air will leak **out** (despite the temperature difference) - actually drying the inside of the wall cavity

0 2 9 9

Differences between old and new construction

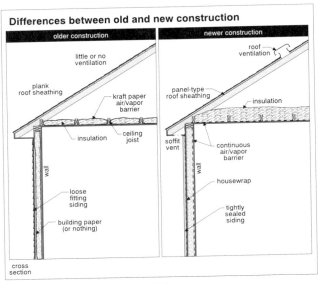

older construction

little or no ventilation

plank roof sheathing

kraft paper air/vapor barrier

insulation

ceiling joist

wall

loose fitting siding

building paper (or nothing)

cross section

newer construction

roof ventilation

panel-type roof sheathing

insulation

soffit vent

continuous air/vapor barrier

wall

housewrap

tightly sealed siding

0 3 0 0

Exhaust fans

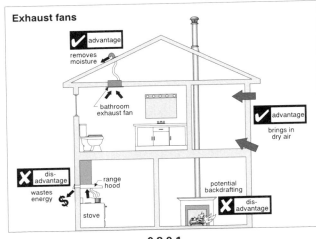

✔ advantage

removes moisture

bathroom exhaust fan

✔ advantage

brings in dry air

✘ dis-advantage

wastes energy

range hood

stove

potential backdrafting

✘ dis-advantage

0 3 0 1

Direct vent appliances

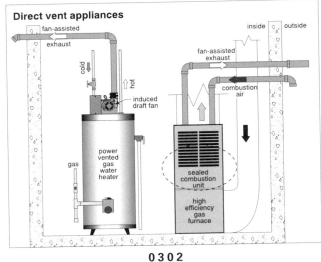

fan-assisted exhaust

cold

hot

induced draft fan

gas

power vented gas water heater

inside | outside

fan-assisted exhaust

combustion air

sealed combustion unit

high efficiency gas furnace

0 3 0 2

Heat recovery ventilator

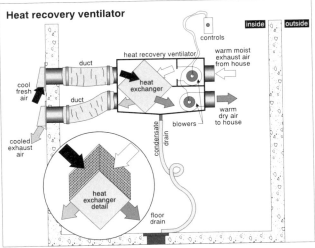

inside | outside

controls

heat recovery ventilator

duct

cool fresh air

duct

heat exchanger

warm moist exhaust air from house

blowers

warm dry air to house

cooled exhaust air

condensate drain

heat exchanger detail

floor drain

0 3 0 3

How depressurization can occur in houses

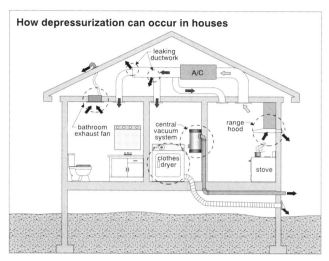

leaking ductwork

A/C

bathroom exhaust fan

central vacuum system

clothes dryer

range hood

stove

0 3 0 4

Controlling moisture in hot climates

A/C

air/vapor barrier should be installed on the outside of the wall or not at all

vinyl wallpaper acts as a vapor barrier and is not recommended as it can cause condensation to be trapped in the wall

don't ventilate crawlspace

crawlspace

dirt floor

gravel

polyethylene sealed at edges

0 3 0 5

Wall marks due to thermal bridging

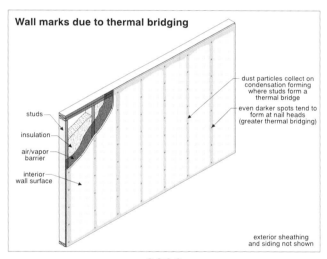

studs

insulation

air/vapor barrier

interior wall surface

dust particles collect on condensation forming where studs form a thermal bridge

even darker spots tend to form at nail heads (greater thermal bridging)

exterior sheathing and siding not shown

0 3 0 6

Insulation voids and convective loops

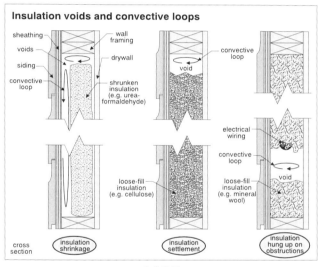

sheathing

voids

siding

convective loop

wall framing

drywall

shrunken insulation (e.g. ureaformaldehyde)

convective loop

void

convective loop

electrical wiring

convective loop

void

loose-fill insulation (e.g. cellulose)

loose-fill insulation (e.g. mineral wool)

cross section

insulation shrinkage

insulation settlement

insulation hung up on obstructions

0 3 0 7

Wind washing

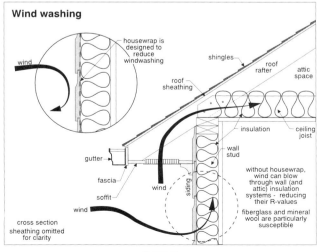

housewrap is designed to reduce windwashing

wind

shingles

roof rafter

attic space

roof sheathing

insulation

ceiling joist

wall stud

without housewrap, wind can blow through wall (and attic) insulation systems - reducing their R-values

fiberglass and mineral wool are particularly susceptible

gutter

fascia

soffit

wind

wind

siding

cross section sheathing omitted for clarity

0 3 0 8

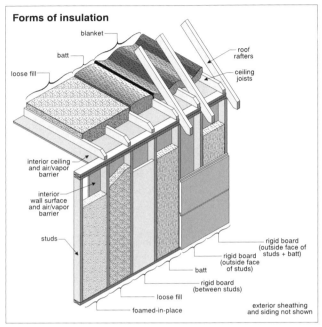

Forms of insulation

blanket
batt
loose fill
roof rafters
ceiling joists
interior ceiling and air/vapor barrier
interior wall surface and air/vapor barrier
studs
rigid board (outside face of studs + batt)
rigid board (outside face of studs)
batt
rigid board (between studs)
loose fill
foamed-in-place
exterior sheathing and siding not shown

0309

Other types of polystyrene insulation

permanent polystyrene forms
poured concrete
bridging between polystyrene forms
plywood
polystyrene

for concrete walls (above and below grade)

prefabricated wall panels

0310

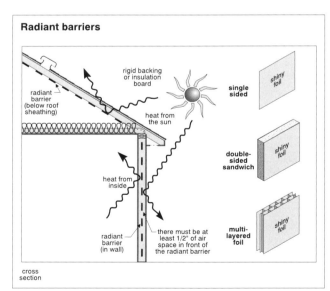

Radiant barriers

rigid backing or insulation board
radiant barrier (below roof sheathing)
heat from the sun
heat from inside
radiant barrier (in wall)
there must be at least 1/2" of air space in front of the radiant barrier
cross section

single sided — shiny foil
double-sided sandwich — shiny foil
multi-layered foil — shiny foil

0311

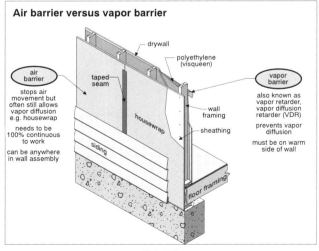

Air barrier versus vapor barrier

drywall
polyethylene (visqueen)
air barrier
taped seam
vapor barrier
wall framing
housewrap
sheathing
siding
floor framing

air barrier
stops air movement but often still allows vapor diffusion e.g. housewrap
needs to be 100% continuous to work
can be anywhere in wall assembly

vapor barrier
also known as vapor retarder, vapor diffusion retarder (VDR)
prevents vapor diffusion
must be on warm side of wall

0312

Housewrap versus building paper

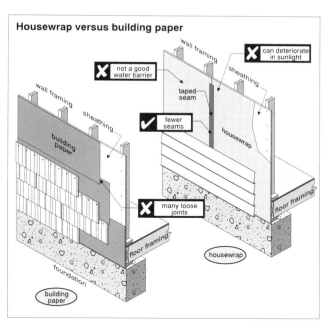

wall framing

not a good water barrier

can deteriorate in sunlight

taped seam

sheathing

wall framing

sheathing

fewer seams

building paper

housewrap

many loose joints

floor framing

floor framing

foundation

housewrap

building paper

0 3 1 3

Sill gaskets and electrical box enclosures

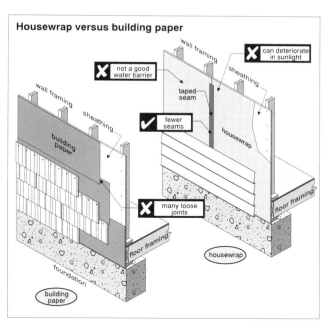

manufactured enclosure sealed to air barrier

sill gasket

sealant

wall framing

air barrier

electrical box enclosure made on site (of polyethylene) and sealed to air barrier

sheathing

foundation

floor framing

insulation, drywall not shown

0 3 1 4

Gaskets for electrical boxes

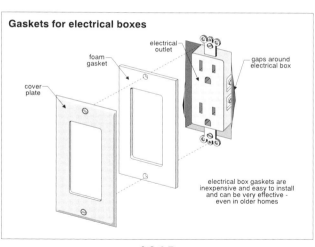

electrical outlet

foam gasket

gaps around electrical box

cover plate

electrical box gaskets are inexpensive and easy to install and can be very effective - even in older homes

0 3 1 5

Backer rods

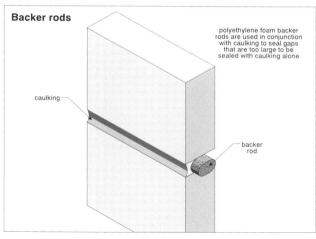

polyethylene foam backer rods are used in conjunction with caulking to seal gaps that are too large to be sealed with caulking alone

caulking

backer rod

0 3 1 6

Caulking - indoors or out?

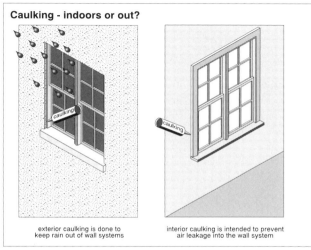

caulking

caulking

exterior caulking is done to keep rain out of wall systems

interior caulking is intended to prevent air leakage into the wall system

0 3 1 7

Pot lights

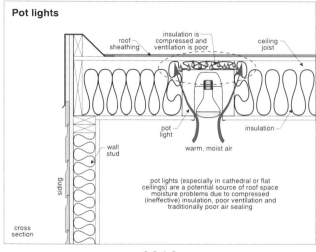

roof sheathing

insulation is compressed and ventilation is poor

ceiling joist

pot light

insulation

wall stud

warm, moist air

siding

pot lights (especially in cathedral or flat ceilings) are a potential source of roof space moisture problems due to compressed (ineffective) insulation, poor ventilation and traditionally poor air sealing

cross section

0 3 1 8

Vapor barrier location

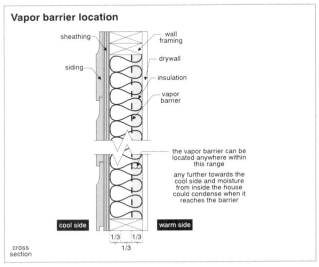

0319

Cover crawlspace floor

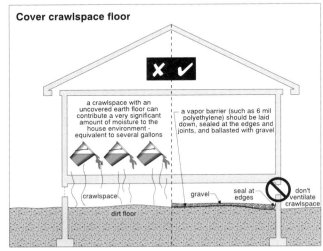

0320

Reducing attic heat with ventilation

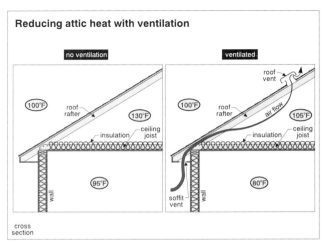

0321

Preventing ice dams with ventilation

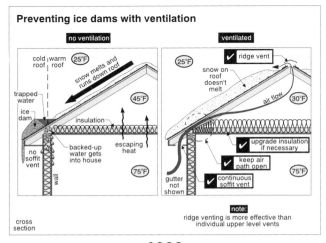

0322

Types and locations of vents

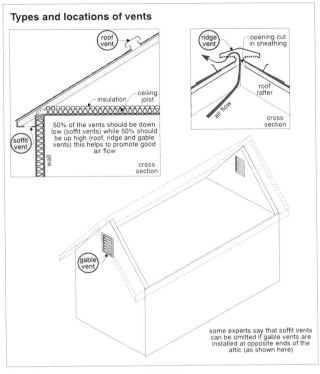

roof vent

ridge vent — opening cut in sheathing

roof rafter

air flow

cross section

insulation — ceiling joist

soffit vent — wall

50% of the vents should be down low (soffit vents) while 50% should be up high (roof, ridge and gable vents) this helps to promote good air flow

cross section

gable vent

some experts say that soffit vents can be omitted if gable vents are installed at opposite ends of the attic (as shown here)

0323

Turbine vents

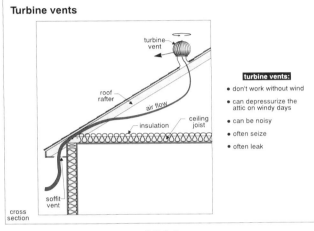

turbine vent

roof rafter

air flow

insulation — ceiling joist

soffit vent

cross section

turbine vents:

- don't work without wind
- can depressurize the attic on windy days
- can be noisy
- often seize
- often leak

0324

Baffles for soffit vents

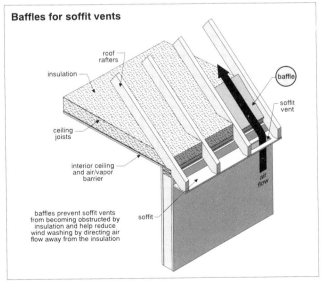

roof rafters

insulation

baffle

soffit vent

ceiling joists

interior ceiling and air/vapor barrier

air flow

soffit

baffles prevent soffit vents from becoming obstructed by insulation and help reduce wind washing by directing air flow away from the insulation

0325

Recommended amount of attic ventilation

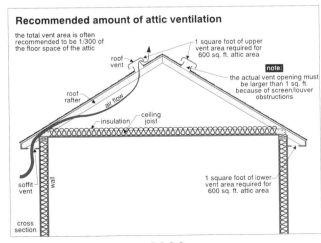

the total vent area is often recommended to be 1/300 of the floor space of the attic

roof vent

1 square foot of upper vent area required for 600 sq. ft. attic area

note: the actual vent opening must be larger than 1 sq. ft. because of screen/louver obstructions

roof rafter

air flow

insulation — ceiling joist

soffit vent — wall

1 square foot of lower vent area required for 600 sq. ft. attic area

cross section

0326

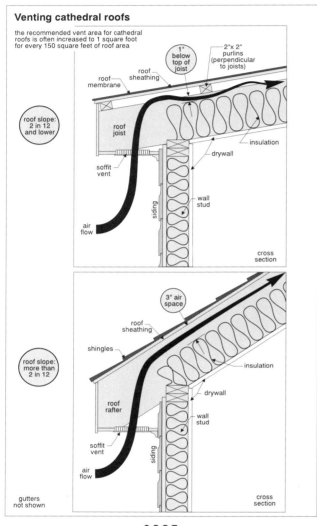

Venting cathedral roofs

the recommended vent area for cathedral roofs is often increased to 1 square foot for every 150 square feet of roof area

roof slope: 2 in 12 and lower

1" below top of joist

2"x 2" purlins (perpendicular to joists)

roof membrane
roof sheathing
roof joist
insulation
drywall
soffit vent
wall stud
siding
air flow

cross section

roof slope: more than 2 in 12

3" air space

roof sheathing
shingles
insulation
drywall
roof rafter
wall stud
soffit vent
siding
air flow

gutters not shown

cross section

0327

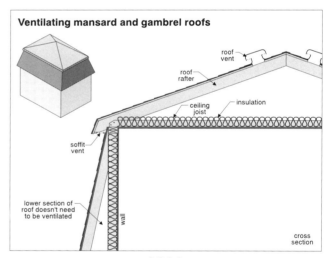

Ventilating mansard and gambrel roofs

roof vent
roof rafter
ceiling joist
insulation
soffit vent
lower section of roof doesn't need to be ventilated
wall

cross section

0328

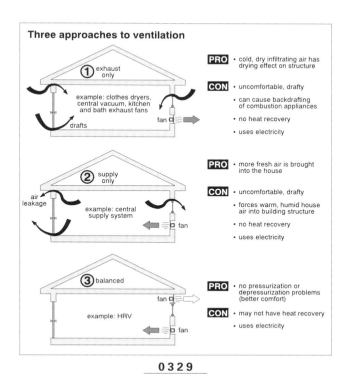

Three approaches to ventilation

(1) exhaust only

example: clothes dryers, central vacuum, kitchen and bath exhaust fans

drafts

fan

PRO
• cold, dry infiltrating air has drying effect on structure

CON
• uncomfortable, drafty
• can cause backdrafting of combustion appliances
• no heat recovery
• uses electricity

(2) supply only

air leakage

example: central supply system

fan

PRO
• more fresh air is brought into the house

CON
• uncomfortable, drafty
• forces warm, humid house air into building structure
• no heat recovery
• uses electricity

(3) balanced

fan

example: HRV

fan

PRO
• no pressurization or depressurization problems (better comfort)

CON
• may not have heat recovery
• uses electricity

0329

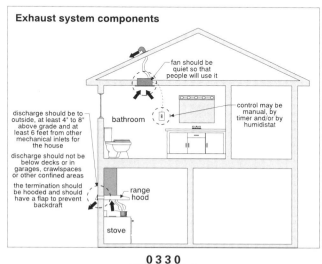

Exhaust system components

fan should be quiet so that people will use it

control may be manual, by timer and/or by humidistat

bathroom

discharge should be to outside, at least 4" to 8" above grade and at least 6 feet from other mechanical inlets for the house

discharge should not be below decks or in garages, crawlspaces or other confined areas

the termination should be hooded and should have a flap to prevent backdraft

range hood

stove

0330

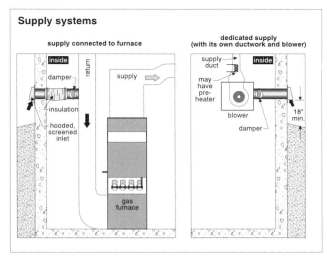

Supply systems

supply connected to furnace

inside

damper

return

supply

insulation

hooded, screened inlet

gas furnace

dedicated supply
(with its own ductwork and blower)

supply duct

inside

may have pre-heater

blower

damper

18" min.

0331

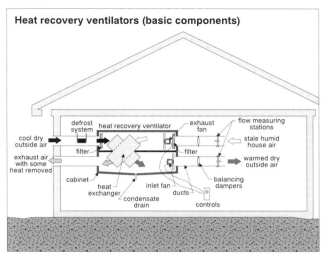

Heat recovery ventilators (basic components)

defrost system

heat recovery ventilator

exhaust fan

flow measuring stations

cool dry outside air

filter

stale humid house air

exhaust air with some heat removed

filter

warmed dry outside air

cabinet

heat exchanger

condensate drain

inlet fan

ducts

controls

balancing dampers

0332

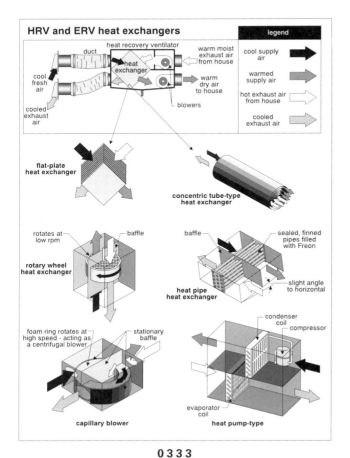

HRV and ERV heat exchangers

legend

heat recovery ventilator
duct
warm moist exhaust air from house
heat exchanger
cool fresh air
warm dry air to house
cooled exhaust air
blowers

cool supply air
warmed supply air
hot exhaust air from house
cooled exhaust air

flat-plate heat exchanger

concentric tube-type heat exchanger

rotates at low rpm
baffle
baffle
sealed, finned pipes filled with Freon
rotary wheel heat exchanger
heat pipe heat exchanger
slight angle to horizontal

foam ring rotates at high speed - acting as a centrifugal blower
stationary baffle
condenser coil
compressor
capillary blower
evaporator coil
heat pump-type

0 3 3 3

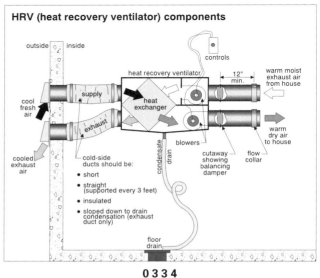

HRV (heat recovery ventilator) components

outside | inside
controls
heat recovery ventilator
12" min.
warm moist exhaust air from house
cool fresh air
supply
heat exchanger
exhaust
warm dry air to house
cooled exhaust air
cold-side ducts should be:
• short
• straight (supported every 3 feet)
• insulated
• sloped down to drain condensation (exhaust duct only)
condensate drain
blowers
cutaway showing balancing damper
flow collar
floor drain

0 3 3 4

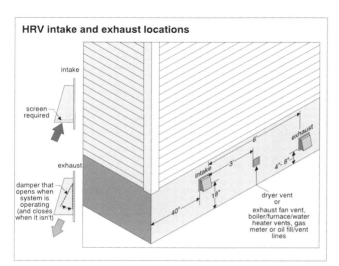

HRV intake and exhaust locations

intake
screen required
exhaust
damper that opens when system is operating (and closes when it isn't)
intake
3'
6'
exhaust
4" - 8"
18"
40"
dryer vent or exhaust fan vent, boiler/furnace/water heater vents, gas meter or oil fill/vent lines

0 3 3 5

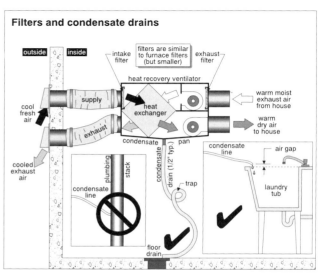

Filters and condensate drains

outside | inside
intake filter
filters are similar to furnace filters (but smaller)
exhaust filter
heat recovery ventilator
cool fresh air
supply
heat exchanger
exhaust
warm moist exhaust air from house
warm dry air to house
condensate pan
cooled exhaust air
plumbing stack
condensate line
condensate drain (1/2" typ.)
trap
condensate line
air gap
laundry tub
floor drain

0 3 3 6

Defrost control

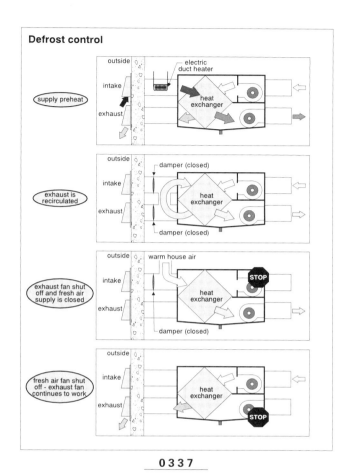

- supply preheat
- exhaust is recirculated
- exhaust fan shut off and fresh air supply is closed
- fresh air fan shut off - exhaust fan continues to work

outside / intake / exhaust / electric duct heater / heat exchanger / damper (closed) / warm house air / STOP

0337

Attic access hatch

the illustration shows a good attic access hatch design

hatches in many houses (especially older ones) won't meet these ideals

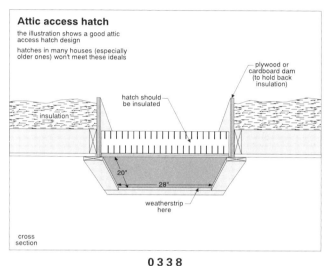

plywood or cardboard dam (to hold back insulation)

hatch should be insulated

insulation

20" / 28"

weatherstrip here

cross section

0338

Stairwells to attic

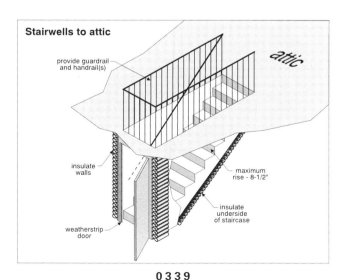

attic

provide guardrail and handrail(s)

insulate walls

weatherstrip door

maximum rise - 8-1/2"

insulate underside of staircase

0339

Pull-down stairs

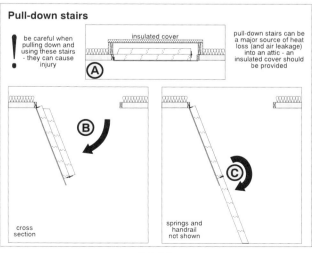

! be careful when pulling down and using these stairs - they can cause injury

insulated cover

(A)

pull-down stairs can be a major source of heat loss (and air leakage) into an attic - an insulated cover should be provided

(B)

(C)

cross section

springs and handrail not shown

0340

Secondary attics

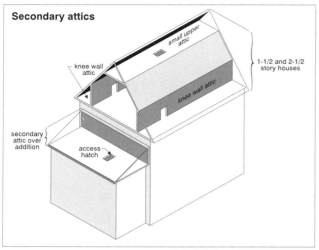

small upper attic

knee wall attic

knee wall attic

1-1/2 and 2-1/2 story houses

secondary attic over addition

access hatch

0 3 4 1

Risks of adding more insulation

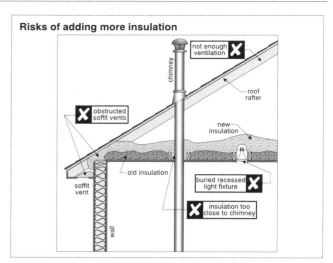

not enough ventilation ✗

chimney

roof rafter

obstructed soffit vents ✗

new insulation

old insulation

soffit vent

wall

buried recessed light fixture ✗

insulation too close to chimney ✗

0 3 4 2

Wet insulation below vents

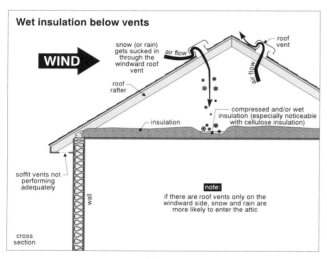

WIND

snow (or rain) gets sucked in through the windward roof vent

air flow

roof vent

air flow

roof rafter

insulation

compressed and/or wet insulation (especially noticeable with cellulose insulation)

soffit vents not performing adequately

wall

note:
if there are roof vents only on the windward side, snow and rain are more likely to enter the attic

cross section

0 3 4 3

Missing insulation at dropped ceiling

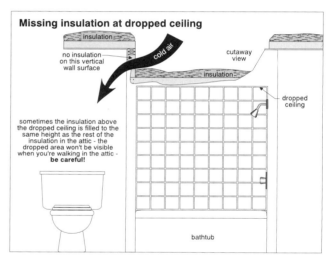

insulation

no insulation on this vertical wall surface

cold air

cutaway view

insulation

dropped ceiling

sometimes the insulation above the dropped ceiling is filled to the same height as the rest of the insulation in the attic - the dropped area won't be visible when you're walking in the attic - **be careful!**

bathtub

0 3 4 4

Insulating knee walls

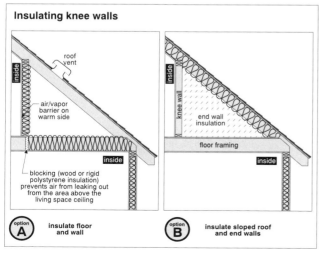

inside

roof vent

air/vapor barrier on warm side

inside

blocking (wood or rigid polystyrene insulation) prevents air from leaking out from the area above the living space ceiling

inside

knee wall

end wall insulation

floor framing

inside

A option insulate floor and wall

B option insulate sloped roof and end walls

0 3 4 5

Insulating skylight wells

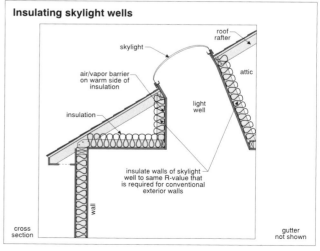

roof rafter

skylight

air/vapor barrier on warm side of insulation

attic

insulation

light well

insulate walls of skylight well to same R-value that is required for conventional exterior walls

wall

cross section

gutter not shown

0 3 4 6

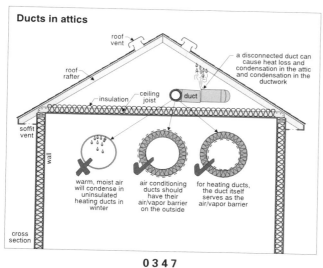

Ducts in attics

roof vent

roof rafter

insulation

ceiling joist

duct

a disconnected duct can cause heat loss and condensation in the attic and condensation in the ductwork

soffit vent

wall

warm, moist air will condense in uninsulated heating ducts in winter

air conditioning ducts should have their air/vapor barrier on the outside

for heating ducts, the duct itself serves as the air/vapor barrier

cross section

0 3 4 7

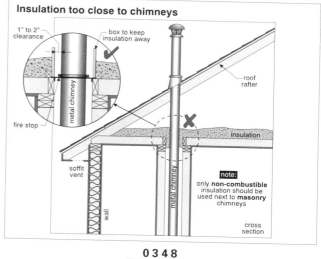

Insulation too close to chimneys

1" to 2" clearance

box to keep insulation away

metal chimney

fire stop

roof rafter

insulation

soffit vent

wall

metal chimney

note: only **non-combustible** insulation should be used next to **masonry** chimneys

cross section

0 3 4 8

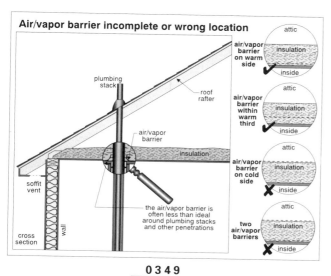

Air/vapor barrier incomplete or wrong location

plumbing stack

roof rafter

air/vapor barrier

insulation

soffit vent

wall

the air/vapor barrier is often less than ideal around plumbing stacks and other penetrations

attic / air/vapor barrier on warm side / insulation / inside

attic / air/vapor barrier within warm third / insulation / inside

attic / air/vapor barrier on cold side / insulation / inside

attic / two air/vapor barriers / insulation / inside

cross section

0 3 4 9

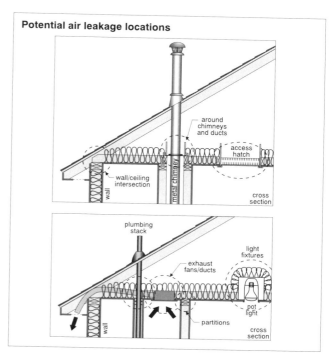

Potential air leakage locations

around chimneys and ducts

access hatch

wall/ceiling intersection

metal chimney

wall

cross section

plumbing stack

light fixtures

exhaust fans/ducts

pot light

partitions

wall

cross section

0 3 5 0

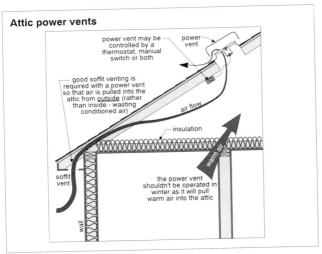

Attic power vents

power vent may be controlled by a thermostat, manual switch or both

power vent

good soffit venting is required with a power vent so that air is pulled into the attic from <u>outside</u> (rather than inside - wasting conditioned air)

air flow

insulation

warm air

soffit vent

the power vent shouldn't be operated in winter as it will pull warm air into the attic

wall

0 3 5 1

Delaminating sheathing

rafter

H-clip

sheathing

rafter

cross section

sheathing

provide 1/8" gap

fasteners can pull through the delaminated sheathing

an H-clip here (or blocking, etc.) helps to prevent panel buckling

sheathing

cross section

install H-clips (or blocking) between rafters/trusses or use tongue and groove sheathing

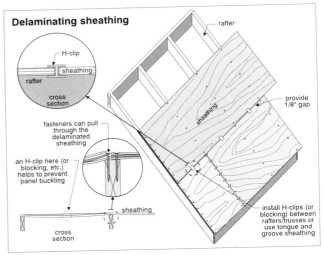

0 3 5 2

Whole house fan with no insulating cover

roof vent

without a cover, a whole house fan can allow air leakage and heat loss into the attic

attic

insulated cover

insulation

whole house fan

soffit vent

wall

cross section

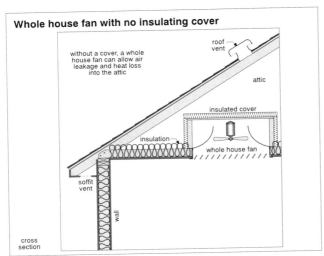

0 3 5 3

Insulating flat (and cathedral) roofs

1 treat as an attic - ventilate above insulation

roof sheathing

roof joist

ventilation space

vent

insulation

siding

wall stud

cross section

2 completely fill roof space

roof sheathing

insulation

siding

wall stud

cross section

3 insulate above roof structure and around perimeter

a insulation above sheathing **b** IRMA*

membrane

ballast

insulation

insulation

insulation

roof joist

roof sheathing

siding

wall stud

cross section

4 insulate below roof structure (retrofit)

roof sheathing

roof joist

vent

ventilation space

insulation

siding

wall stud

original ceiling

new drywall

cross section

* Inverted Roof Membrane Assembly roof or protected membrane roof

0 3 5 4

Evidence of insulation added

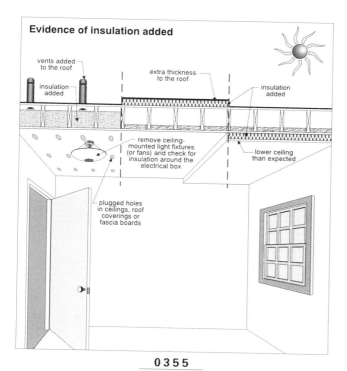

vents added to the roof

insulation added

extra thickness to the roof

insulation added

remove ceiling-mounted light fixtures (or fans) and check for insulation around the electrical box

lower ceiling than expected

plugged holes in ceilings, roof coverings or fascia boards

0355

Two different strategies for insulating cathedral/flat roofs

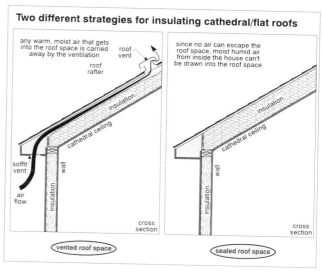

any warm, moist air that gets into the roof space is carried away by the ventilation

roof vent

roof rafter

insulation

cathedral ceiling

soffit vent

air flow

insulation

wall

cross section

vented roof space

since no air can escape the roof space, moist humid air from inside the house can't be drawn into the roof space

insulation

cathedral ceiling

insulation

wall

cross section

sealed roof space

0356

Channeled vents versus cross ventilation

for cathedral ceilings and flat roofs the recommended vent area is 1 square foot for every 150 square feet of roof area

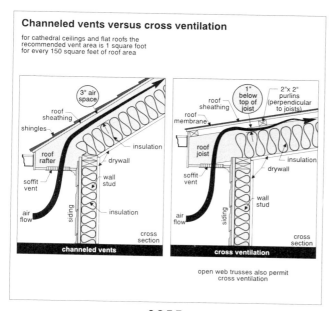

3" air space

roof sheathing

shingles

roof rafter

soffit vent

air flow

insulation

drywall

wall stud

siding

insulation

cross section

channeled vents

1" below top of joist

2"x 2" purlins (perpendicular to joists)

roof sheathing

roof membrane

roof joist

soffit vent

air flow

insulation

drywall

wall stud

siding

insulation

cross section

cross ventilation

open web trusses also permit cross ventilation

0357

Insulation short circuit

insulation is sometimes added to the exterior of a wood frame wall without adding insulation in the stud cavity

this can be a waste of time because convective currents circulate through the empty stud cavity and carry heat out the top of the stud wall

it's kind of like wearing a hat a foot above your head

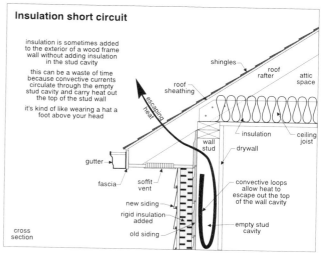

shingles

roof rafter

attic space

roof sheathing

escaping heat

gutter

fascia

soffit vent

new siding

rigid insulation added

old siding

wall stud

insulation

ceiling joist

drywall

convective loops allow heat to escape out the top of the wall cavity

empty stud cavity

cross section

0358

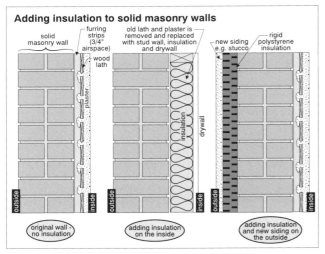

Adding insulation to solid masonry walls

solid masonry wall

furring strips (3/4" airspace)

wood lath

plaster

old lath and plaster is removed and replaced with stud wall, insulation and drywall

insulation

drywall

new siding e.g. stucco

rigid polystyrene insulation

outside

inside

original wall - no insulation

adding insulation on the inside

adding insulation and new siding on the outside

0359

Cold wall effect

even if the room temperature is a comfortable 72°F, a cold wall can draw heat from your body, making you feel uncomfortable

room temperature 72°F

heat flow

cold, uninsulated wall

0360

Drying potential of walls

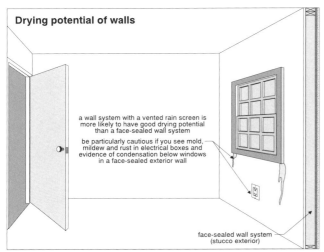

a wall system with a vented rain screen is more likely to have good drying potential than a face-sealed wall system

be particularly cautious if you see mold, mildew and rust in electrical boxes and evidence of condensation below windows in a face-sealed exterior wall

face-sealed wall system (stucco exterior)

0361

Adding exterior basement insulation

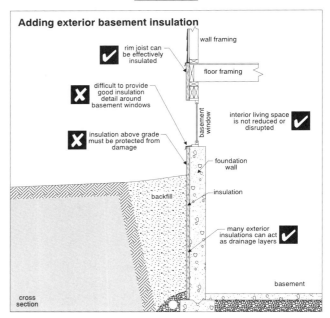

rim joist can be effectively insulated

wall framing

floor framing

difficult to provide good insulation detail around basement windows

insulation above grade must be protected from damage

basement window

interior living space is not reduced or disrupted

foundation wall

insulation

backfill

many exterior insulations can act as drainage layers

basement

cross section

0362

Adding interior basement insulation

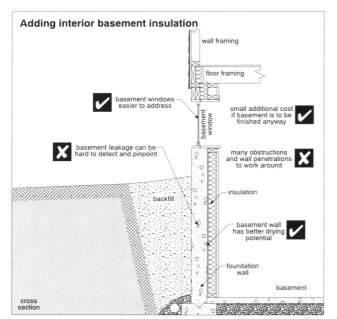

wall framing

floor framing

basement windows easier to address

small additional cost if basement is to be finished anyway

basement window

basement leakage can be hard to detect and pinpoint

many obstructions and wall penetrations to work around

insulation

backfill

basement wall has better drying potential

foundation wall

basement

cross section

0363

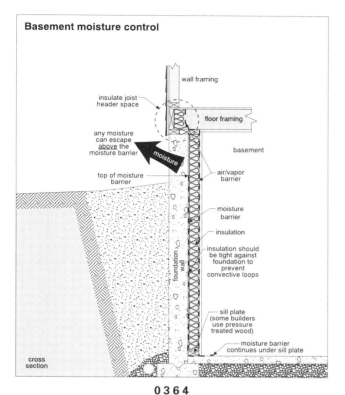

Basement moisture control

- wall framing
- insulate joist header space
- floor framing
- any moisture can escape <u>above</u> the moisture barrier
- moisture
- basement
- top of moisture barrier
- air/vapor barrier
- moisture barrier
- insulation
- insulation should be tight against foundation to prevent convective loops
- foundation wall
- sill plate (some builders use pressure treated wood)
- moisture barrier continues under sill plate
- cross section

0 3 6 4

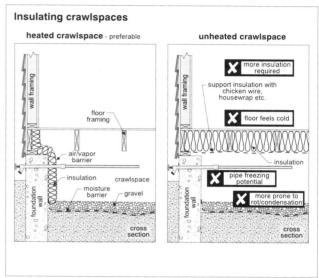

Insulating crawlspaces

heated crawlspace - preferable

- wall framing
- floor framing
- air/vapor barrier
- insulation
- crawlspace
- foundation wall
- moisture barrier
- gravel
- cross section

unheated crawlspace

- wall framing
- ✗ more insulation required
- support insulation with chicken wire, housewrap etc.
- ✗ floor feels cold
- insulation
- ✗ pipe freezing potential
- ✗ more prone to rot/condensation
- foundation wall
- cross section

0 3 6 5

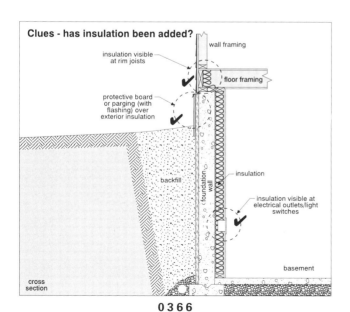

Clues - has insulation been added?

- wall framing
- insulation visible at rim joists
- floor framing
- protective board or parging (with flashing) over exterior insulation
- backfill
- foundation wall
- insulation
- insulation visible at electrical outlets/light switches
- basement
- cross section

0 3 6 6

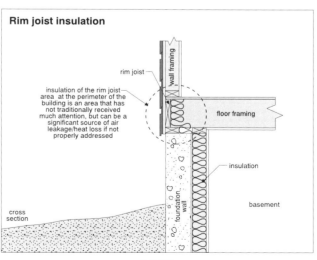

Rim joist insulation

- rim joist
- wall framing
- insulation of the rim joist area at the perimeter of the building is an area that has not traditionally received much attention, but can be a significant source of air leakage/heat loss if not properly addressed
- floor framing
- insulation
- foundation wall
- basement
- cross section

0 3 6 7

Don't insulate embedded joists

insulating around embedded joists can lead to rotting

good air sealing with caulking is a better alternative

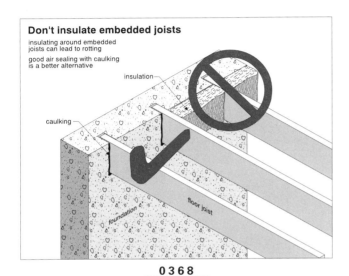

insulation

caulking

foundation

floor joist

0368

Exposed combustible insulation

combustible insulation in basements, crawlspaces, garages or any other areas exposed to people, heating and/or electricity sources should be covered with drywall (wood paneling is acceptable in most areas)

✗ exposed plastic insulation on garage door, walls and ceiling	✔ insulation removed or covered with noncombustible material (drywall)

plastic insulation

plastic insulation
plastic insulation
plastic insulation
plastic insulation
plastic insulation

remove plastic insulation

cover plastic insulation on walls and ceilings

0369

Cover crawlspace floor

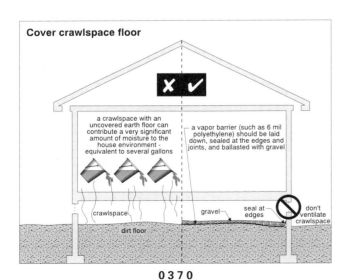

a crawlspace with an uncovered earth floor can contribute a very significant amount of moisture to the house environment - equivalent to several gallons

a vapor barrier (such as 6 mil polyethylene) should be laid down, sealed at the edges and joints, and ballasted with gravel

crawlspace

dirt floor

gravel

seal at edges

don't ventilate crawlspace

0370

Insulation in floors over unheated areas

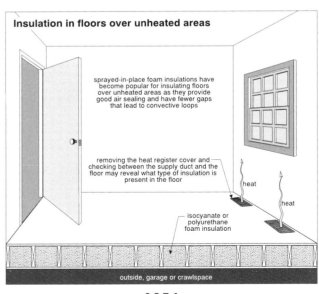

sprayed-in-place foam insulations have become popular for insulating floors over unheated areas as they provide good air sealing and have fewer gaps that lead to convective loops

removing the heat register cover and checking between the supply duct and the floor may reveal what type of insulation is present in the floor

heat

heat

isocyanate or polyurethane foam insulation

outside, garage or crawlspace

0371

Reducing moisture

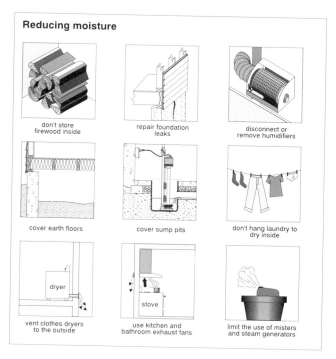

don't store firewood inside

repair foundation leaks

disconnect or remove humidifiers

cover earth floors

cover sump pits

don't hang laundry to dry inside

vent clothes dryers to the outside

use kitchen and bathroom exhaust fans

limit the use of misters and steam generators

0372

Exhaust fan conditions

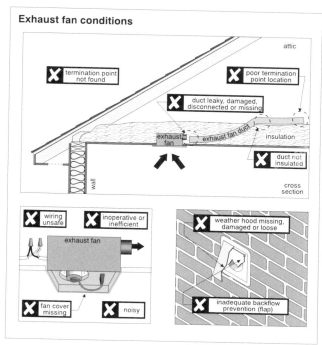

attic

✗ termination point not found

✗ poor termination point location

✗ duct leaky, damaged, disconnected or missing

exhaust fan

exhaust fan duct

insulation

✗ duct not insulated

wall

cross section

✗ wiring unsafe

✗ inoperative or inefficient

exhaust fan

✗ fan cover missing

✗ noisy

✗ weather hood missing, damaged or loose

✗ inadequate backflow prevention (flap)

0373

Cold-side ducts not insulated

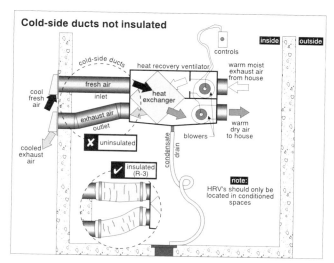

inside / outside

controls

cold-side ducts

heat recovery ventilator

warm moist exhaust air from house

fresh air

cool fresh air

inlet

heat exchanger

exhaust air

outlet

cooled exhaust air

✗ uninsulated

blowers

condensate drain

warm dry air to house

✓ insulated (R-3)

note:
HRV's should only be located in conditioned spaces

0374

Furnace duct/HRV connection

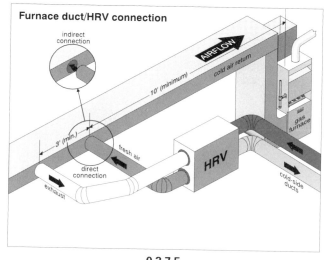

indirect connection

AIRFLOW

cold air return

10' (minimum)

3' (min.)

fresh air

direct connection

exhaust

gas furnace

HRV

cold-side ducts

0375

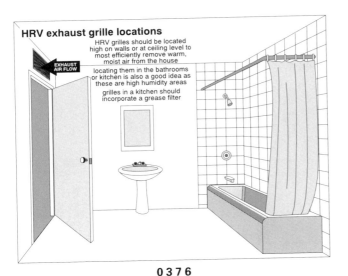

HRV exhaust grille locations

HRV grilles should be located high on walls or at ceiling level to most efficiently remove warm, moist air from the house

locating them in the bathrooms or kitchen is also a good idea as these are high humidity areas

grilles in a kitchen should incorporate a grease filter

EXHAUST AIR FLOW

0 3 7 6

Flow collars and balancing dampers

outside | inside

heat recovery ventilator

cool fresh air

supply

heat exchanger

blower

blower

12" min.

warm moist exhaust air from house

warm dry air to house

condensate drain

cutaway showing balancing damper

cooled exhaust air

flow collar

the flow collar can be identified by the two large pins projecting from the collar - these pins are the connection point for gauges that measure airflow (to balance the HRV)

flow collar

0 3 7 7

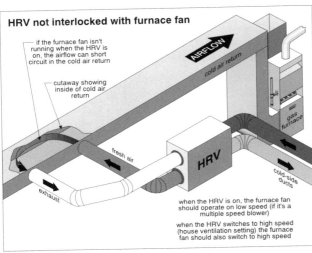

HRV not interlocked with furnace fan

if the furnace fan isn't running when the HRV is on, the airflow can short circuit in the cold air return

cutaway showing inside of cold air return

AIRFLOW

cold air return

gas furnace

fresh air

HRV

exhaust

cold-side ducts

when the HRV is on, the furnace fan should operate on low speed (if it's a multiple speed blower)

when the HRV switches to high speed (house ventilation setting) the furnace fan should also switch to high speed

0 3 7 8

Filters dirty or missing

outside | inside

intake filter

heat recovery ventilator

exhaust filter

cool fresh air

supply

heat exchanger

warm moist exhaust air from house

warm dry air to house

exhaust

condensate drain

pan

cooled exhaust air

filters are similar to furnace filters (but smaller)

remove the filters to check their cleanliness (the dirty side of the filter faces away from the heat exchanger)

condensate drain

floor drain

HRV filters are often neglected

0 3 7 9

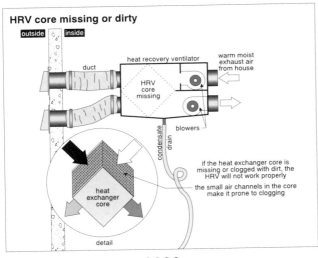

HRV core missing or dirty

outside | inside

duct

heat recovery ventilator

warm moist exhaust air from house

HRV core missing

condensate drain

blowers

if the heat exchanger core is missing or clogged with dirt, the HRV will not work properly

the small air channels in the core make it prone to clogging

heat exchanger core

detail

0 3 8 0

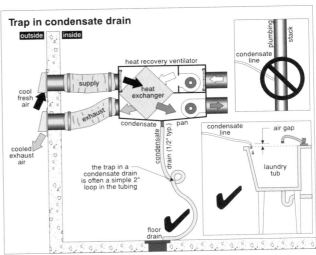

Trap in condensate drain

outside | inside

heat recovery ventilator

cool fresh air

supply

heat exchanger

exhaust

condensate

pan

cooled exhaust air

condensate drain (1/2" typ.)

the trap in a condensate drain is often a simple 2" loop in the tubing

floor drain

plumbing stack

condensate line

condensate line

air gap

laundry tub

0 3 8 1

PART 4

THE INTERIOR

PART 4

INTERIORS

Fire and gas proofing in attached garages

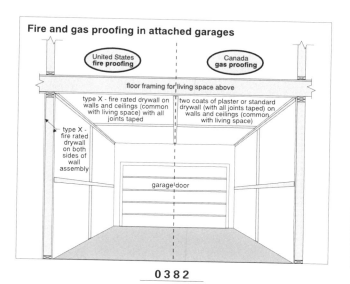

United States fire proofing

Canada gas proofing

floor framing for living space above

type X - fire rated drywall on walls and ceilings (common with living space) with all joints taped

two coats of plaster or standard drywall (with all joints taped) on walls and ceilings (common with living space)

type X - fire rated drywall on both sides of wall assembly

garage door

0382

Control joints in concrete floors

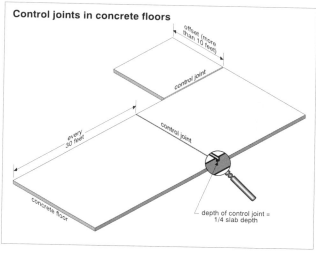

offset (more than 10 feet)

control joint

every 30 feet

control joint

concrete floor

depth of control joint = 1/4 slab depth

0383

Rot around plumbing fixtures

we find more rotted wood subflooring, joists and beams around toilets than any other plumbing fixture (shower stalls come a close second)

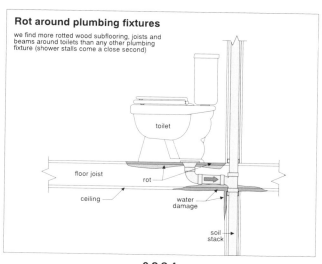

toilet

floor joist

rot

ceiling

water damage

soil stack

0384

Sources of interior water damage

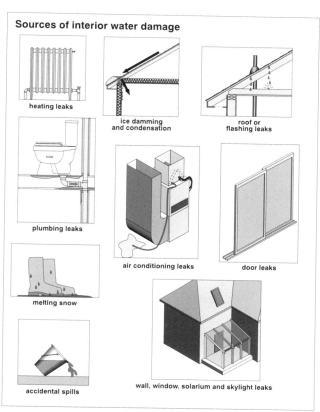

heating leaks

ice damming and condensation

roof or flashing leaks

toilet

plumbing leaks

air conditioning leaks

door leaks

melting snow

accidental spills

wall, window, solarium and skylight leaks

0385

Concrete floor problems

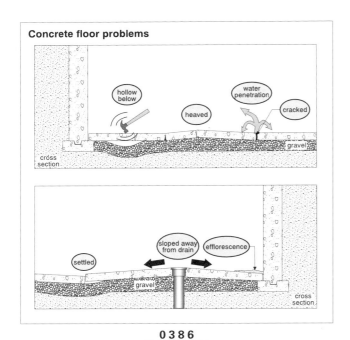

hollow below

heaved

water penetration

cracked

cross section

gravel

settled

sloped away from drain

efflorescence

gravel

cross section

0386

Hollows

you should document any hollow spaces that you find, but you won't be able to determine the size and severity of the voids below the floor

recommend monitoring for localized hollows and further investigation for more extensive ones

hammers or heavy chains are used by specialists to detect hollow spots

foundation wall

floor slab   void   gravel

cross section

0387

Causes of rot

wood is vulnerable to rot attack when the moisture content is above 20%

air must also be present (wood totally submerged in water will rot very slowly, or not at all)

0388

Alternatives for installing ceramic tiles

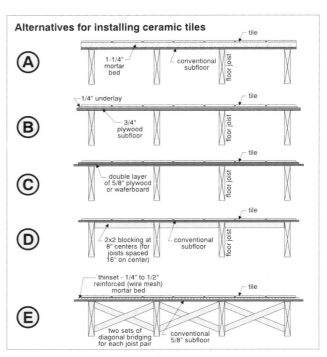

(A)   tile
1-1/4" mortar bed   conventional subfloor   floor joist

(B)   1/4" underlay   tile
3/4" plywood subfloor   floor joist

(C)   tile
double layer of 5/8" plywood or waferboard   floor joist

(D)   tile
2x2 blocking at 8" centers (for joists spaced 16" on center)   conventional subfloor   floor joist

(E)   thinset - 1/4" to 1/2" reinforced (wire mesh) mortar bed   tile
two sets of diagonal bridging for each joist pair   conventional 5/8" subfloor

0389

Gypsum lath versus drywall

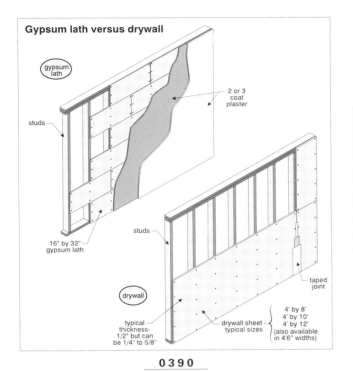

gypsum lath

2 or 3 coat plaster

studs

16" by 32" gypsum lath

studs

drywall

taped joint

typical thickness- 1/2" but can be 1/4" to 5/8"

drywall sheet - typical sizes

4' by 8'
4' by 10'
4' by 12'
(also available in 4'6" widths)

0390

Inspecting walls

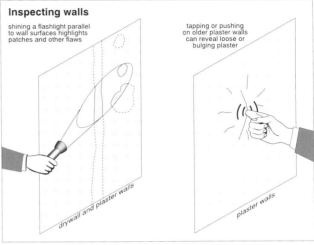

shining a flashlight parallel to wall surfaces highlights patches and other flaws

tapping or pushing on older plaster walls can reveal loose or bulging plaster

drywall and plaster walls

plaster walls

0391

Common locations for water damage

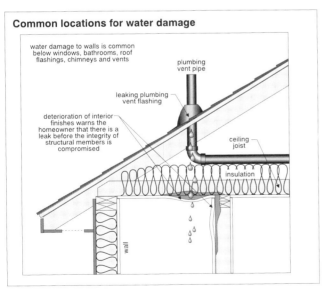

water damage to walls is common below windows, bathrooms, roof flashings, chimneys and vents

plumbing vent pipe

leaking plumbing vent flashing

deterioration of interior finishes warns the homeowner that there is a leak before the integrity of structural members is compromised

ceiling joist

insulation

wall

0392

Structural clues

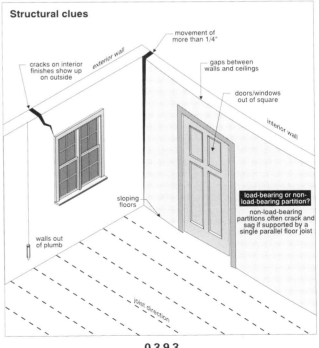

movement of more than 1/4"

cracks on interior finishes show up on outside

exterior wall

gaps between walls and ceilings

doors/windows out of square

interior wall

sloping floors

walls out of plumb

load-bearing or non-load-bearing partition?

non-load-bearing partitions often crack and sag if supported by a single parallel floor joist

joist direction

0393

Shadow effect

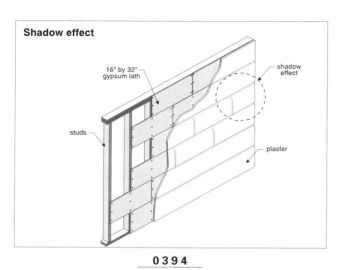

16" by 32" gypsum lath

shadow effect

studs

plaster

0394

Nail pop mechanism

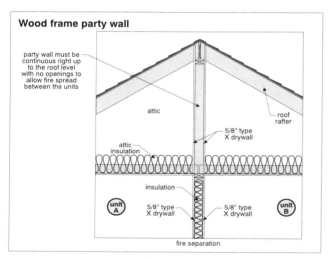

nail

drywall compound

wall stud

drywall

(A) wall stud shrinks as it dries

(B) force against the wall pushes the drywall back against the stud

the nail stays out, popping off the drywall compound over the nail

1-1/4" nails are best for 1/2" drywall

shorter nails are less prone to popping than longer nails

1-1/8" screws are best for 1/2" drywall

screws hold 3 times better than nails, will not pop and are less likely to tear the paper drywall surface

0395

Wood frame party wall

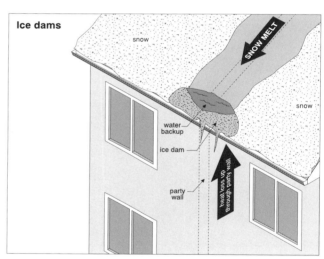

party wall must be continuous right up to the roof level with no openings to allow fire spread between the units

attic

roof rafter

5/8" type X drywall

attic insulation

insulation

unit A

5/8" type X drywall

5/8" type X drywall

unit B

fire separation

0396

Ice dams

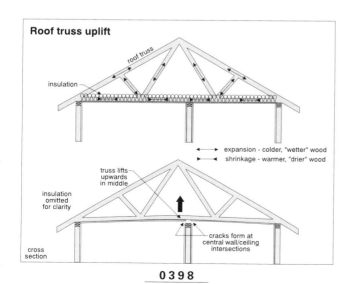

snow

SNOW MELT

water backup

ice dam

snow

heat loss up through party wall

party wall

0397

Roof truss uplift

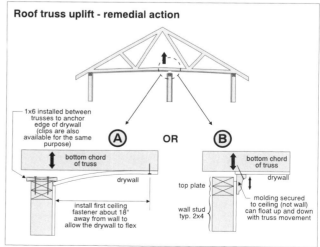

roof truss

insulation

expansion - colder, "wetter" wood
shrinkage - warmer, "drier" wood

truss lifts upwards in middle

insulation omitted for clarity

cracks form at central wall/ceiling intersections

cross section

0398

Roof truss uplift - remedial action

1x6 installed between trusses to anchor edge of drywall (clips are also available for the same purpose)

(A) OR (B)

bottom chord of truss

drywall

install first ceiling fastener about 18" away from wall to allow the drywall to flex

bottom chord of truss

drywall

top plate

molding secured to ceiling (not wall) can float up and down with truss movement

wall stud typ. 2x4

0399

Strapping the underside of trusses

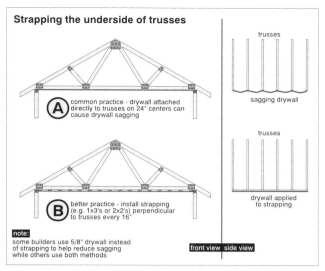

A common practice - drywall attached directly to trusses on 24" centers can cause drywall sagging

B better practice - install strapping (e.g. 1x3's or 2x2's) perpendicular to trusses every 16"

trusses

sagging drywall

trusses

drywall applied to strapping

note: some builders use 5/8" drywall instead of strapping to help reduce sagging while others use both methods

front view side view

0 4 0 0

Stairway lighting

stairway lighting requires switches at both the top and bottom of the stairs when the stairs have more than 3 treads (CAN) or more than 6 treads (USA)

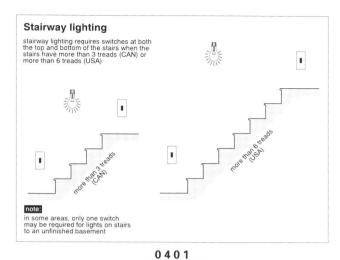

more than 3 treads (CAN)

more than 6 treads (USA)

note: in some areas, only one switch may be required for lights on stairs to an unfinished basement

0 4 0 1

Kinds of trim

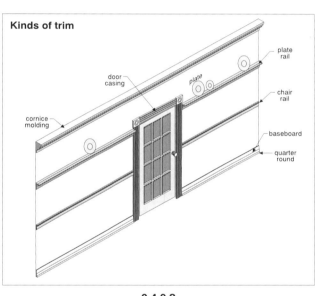

plate rail

door casing

plate

chair rail

cornice molding

baseboard

quarter round

0 4 0 2

Counter problems

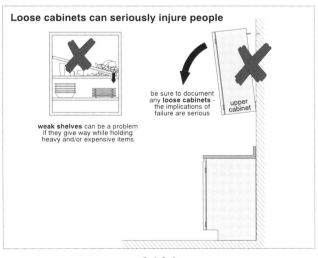

stains or rust

burns

cuts

loose, missing or cracked tiles (or grout) on ceramics

mechanical damage

loose or missing pieces

0 4 0 3

Loose cabinets can seriously injure people

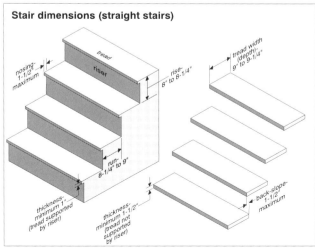

be sure to document any **loose cabinets** - the implications of failure are serious

upper cabinet

weak shelves can be a problem if they give way while holding heavy and/or expensive items

0 4 0 4

Stair dimensions (straight stairs)

tread

riser

nosing- 1-1/2" maximum

rise- 8" to 8-1/4"

tread width (depth)- 9" to 9-1/4"

run- 8-1/4" to 9"

thickness- minimum 1" (tread supported by riser)

thickness- minimum 1-1/2" (tread not supported by riser)

back-slope- 1-1/2" maximum

0 4 0 5

Curved treads

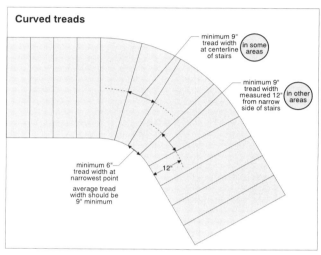

minimum 9"
tread width
at centerline
of stairs — (in some areas)

minimum 9"
tread width
measured 12"
from narrow
side of stairs — (in other areas)

minimum 6"
tread width at
narrowest point

average tread
width should be
9" minimum

12"

0406

Winders

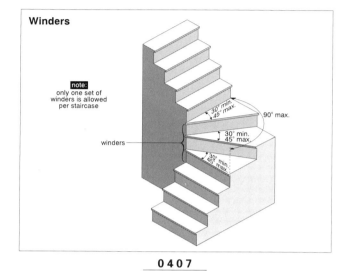

note:
only one set of
winders is allowed
per staircase

winders

30° min.
45° max.

90° max.

30° min.
45° max.

30° min.
45° max.

0407

Stringers

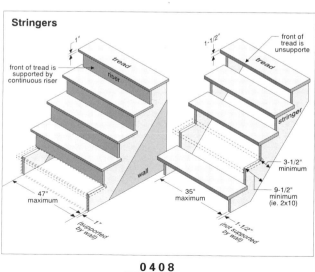

front of tread is
supported by
continuous riser

1"

tread

riser

wall

47"
maximum

1"
(supported
by wall)

1-1/2"

front of
tread is
unsupporte

tread

stringer

3-1/2"
minimum

9-1/2"
minimum
(ie. 2x10)

35"
maximum

1-1/2"
(not supported
by wall)

0408

Stairwell width

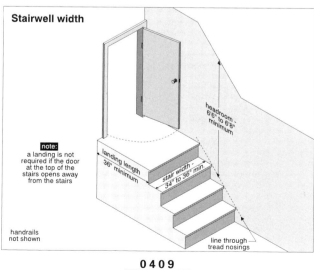

note:
a landing is not
required if the door
at the top of the
stairs opens away
from the stairs

headroom -
6'6" to 6'8"
minimum

landing length -
36" minimum

stair width -
34" to 36" min.

handrails
not shown

line through
tread nosings

0409

Handrails and guards

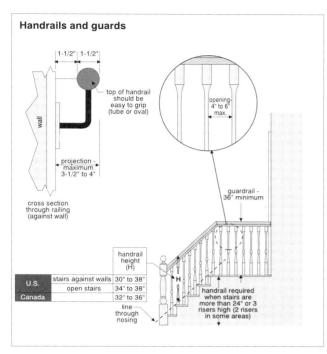

1-1/2" | 1-1/2"

top of handrail should be easy to grip (tube or oval)

wall

projection - maximum 3-1/2" to 4"

cross section through railing (against wall)

opening - 4" to 6" max.

guardrail - 36" minimum

		handrail height (H)
U.S.	stairs against walls	30" to 38"
	open stairs	34" to 38"
Canada		32" to 36"

H

handrail required when stairs are more than 24" or 3 risers high (2 risers in some areas)

line through nosing

0 4 1 0

Rot in wood stairs

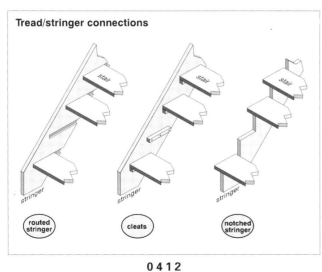

stair stringers against exterior basement walls can be prone to rotting - check these carefully

stringer

exterior wall

watch for rot at the bottom of basement stair stringers (especially if they extend into the basement floor slab)

0 4 1 1

Tread/stringer connections

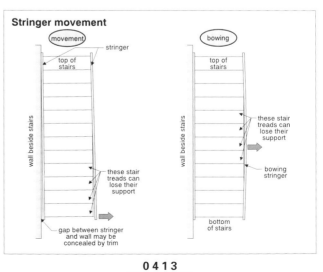

stair

stringer

routed stringer

cleats

notched stringer

0 4 1 2

Stringer movement

movement

stringer

top of stairs

wall beside stairs

these stair treads can lose their support

gap between stringer and wall may be concealed by trim

bowing

top of stairs

wall beside stairs

these stair treads can lose their support

bowing stringer

bottom of stairs

0 4 1 3

Handrail support

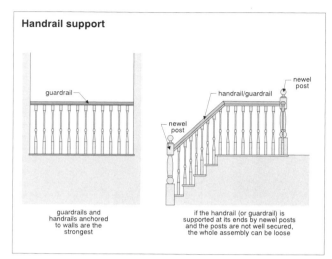

guardrail

handrail/guardrail

newel post

newel post

guardrails and handrails anchored to walls are the strongest

if the handrail (or guardrail) is supported at its ends by newel posts and the posts are not well secured, the whole assembly can be loose

0 4 1 4

Fire stopping for stairs

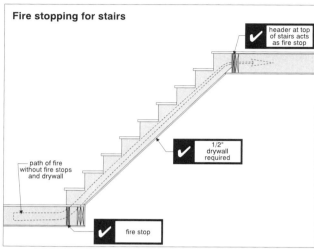

✓ header at top of stairs acts as fire stop

✓ 1/2" drywall required

path of fire without fire stops and drywall

✓ fire stop

0 4 1 5

Glazing types

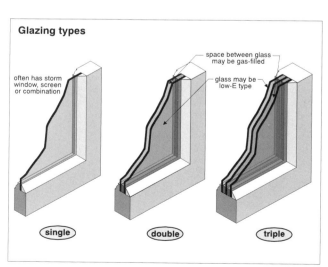

often has storm window, screen or combination

space between glass may be gas-filled

glass may be low-E type

single

double

triple

0 4 1 6

Window types

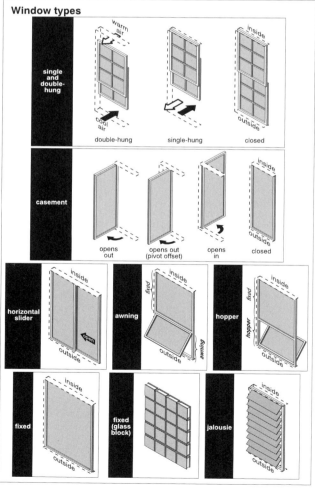

single and double-hung

warm air

cool air

inside

outside

double-hung

single-hung

closed

casement

inside

outside

opens out

opens out (pivot offset)

opens in

closed

horizontal slider

inside

outside

open

awning

fixed

inside

awning

outside

hopper

fixed

inside

hopper

outside

fixed

inside

outside

fixed (glass block)

jalousie

inside

outside

0 4 1 7

Window components

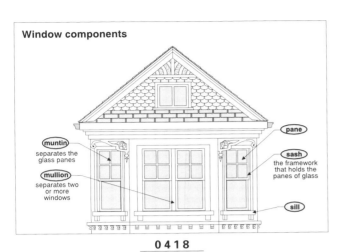

muntin
separates the glass panes

mullion
separates two or more windows

pane

sash
the framework that holds the panes of glass

sill

0418

Fall protection

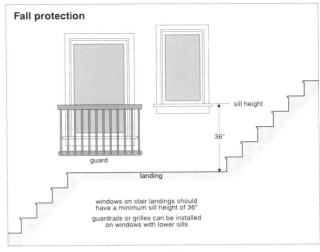

sill height

36"

guard

landing

windows on stair landings should have a minimum sill height of 36"

guardrails or grilles can be installed on windows with lower sills

0419

Drain holes

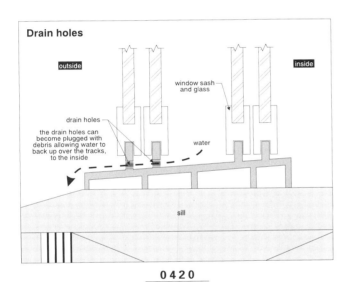

outside

inside

window sash and glass

drain holes

the drain holes can become plugged with debris allowing water to back up over the tracks, to the inside

water

sill

0420

Flashing pan

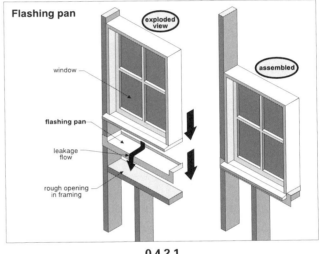

exploded view

assembled

window

flashing pan

leakage flow

rough opening in framing

0421

Energy efficiency

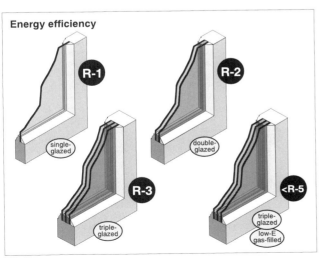

R-1
single-glazed

R-2
double-glazed

R-3
triple-glazed

<R-5
triple-glazed low-E gas-filled

0422

Low-E glass

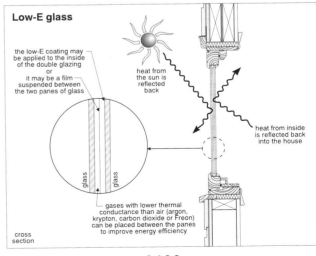

the low-E coating may be applied to the inside of the double glazing
or
it may be a film suspended between the two panes of glass

heat from the sun is reflected back

heat from inside is reflected back into the house

glass

glass

gases with lower thermal conductance than air (argon, krypton, carbon dioxide or Freon) can be placed between the panes to improve energy efficiency

cross section

0423

Move glass inboard

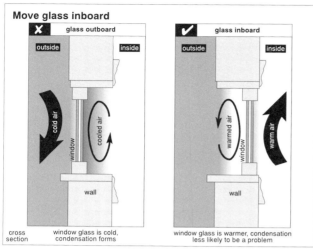

glass outboard ✗
outside / inside
cold air / cooled air / window
wall

cross section
window glass is cold, condensation forms

glass inboard ✓
outside / inside
warmed air / window / warm air
wall

window glass is warmer, condensation less likely to be a problem

0424

Spacing between layers of glass

spacing too small - conductive heat loss ✗
spacing too large - creation of convective loops ✗
spacing correct - about 5/8" ✓
heat loss
outside glass / inside glass

less heat loss
metal spacers have greater thermal conductivity than rubber spacers (let more heat escape)
more heat loss
more heat is lost around the edge of a window (through the sash) than in the middle of the window - this is called **edge effect**

cross section

0425

Thermal breaks in frames

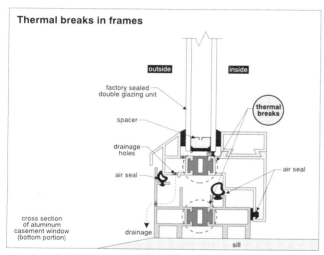

outside / inside
factory sealed double glazing unit
spacer
drainage holes
air seal
thermal breaks
air seal
cross section of aluminum casement window (bottom portion)
drainage
sill

0426

Look behind window treatments

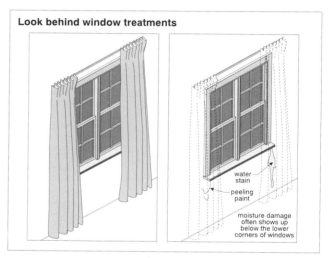

water stain
peeling paint
moisture damage often shows up below the lower corners of windows

0427

Sloped glazing

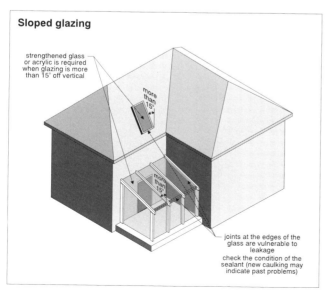

strengthened glass or acrylic is required when glazing is more than 15° off vertical
more than 15°
more than 15°
joints at the edges of the glass are vulnerable to leakage
check the condition of the sealant (new caulking may indicate past problems)

0428

Skylight and solarium leaks

skylights and solariums are very prone to leakage
leakage typically occurs through the flashings or at the edges of the glass

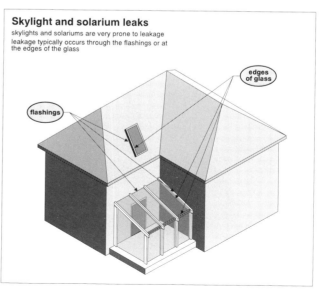

edges of glass
flashings

0429

Lintels sagging or missing

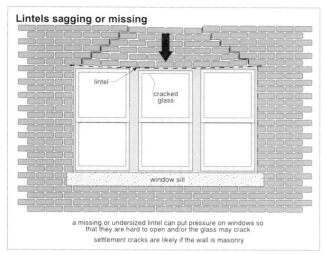

lintel

cracked glass

window sill

a missing or undersized lintel can put pressure on windows so that they are hard to open and/or the glass may crack

settlement cracks are likely if the wall is masonry

0430

Window frame deformation

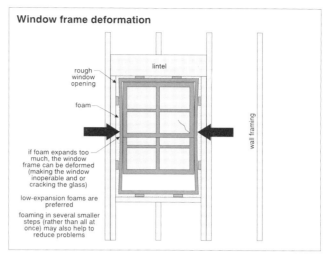

rough window opening

lintel

foam

wall framing

if foam expands too much, the window frame can be deformed (making the window inoperable and or cracking the glass)

low-expansion foams are preferred

foaming in several smaller steps (rather than all at once) may also help to reduce problems

0431

Window installed backwards

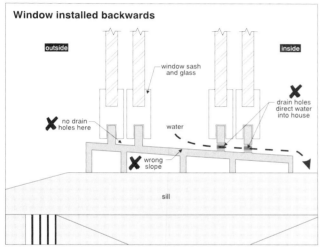

outside

inside

window sash and glass

no drain holes here

drain holes direct water into house

water

wrong slope

sill

0432

Flashings over windows

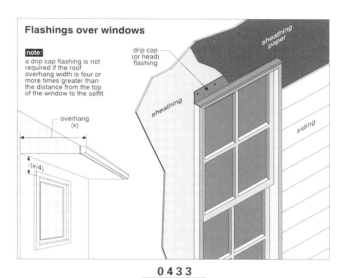

note:
a drip cap flashing is not required if the roof overhang width is four or more times greater than the distance from the top of the window to the soffit

drip cap (or head) flashing

sheathing paper

sheathing

siding

overhang (x)

(x/4)

0433

Watch for faulty windows

be careful when operating double-hung and single-hung windows (and also self-storing storms)

if they're defective, they could fall unexpectedly - injuring your hands and/or the window

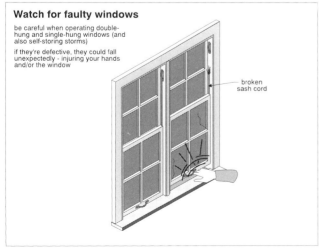

broken sash cord

0434

Don't push on upper sash

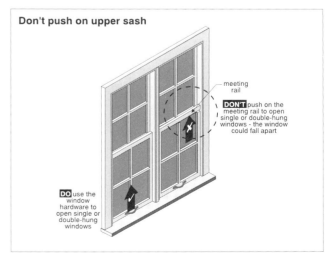

meeting rail

DON'T push on the meeting rail to open single or double-hung windows - the window could fall apart

DO use the window hardware to open single or double-hung windows

0435

Exit window sizes

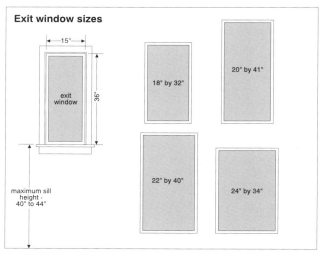

exit window

15"

36"

18" by 32"

20" by 41"

22" by 40"

24" by 34"

maximum sill height - 40" to 44"

0436

Ice damming below skylight

problem

localized heat loss causes snow to melt around the skylight

the water refreezes when it hits the colder roof below - building up a dam

water running down the roof can back up under the shingles or skylight flashings

snow

skylight

water buildup

ice dam

attic

(insulation omitted for clarity)

heat loss

roof rafter

snow

water entry

ceiling joist

exterior wall

solution

install Ice and Water Shield below the shingles for 6' around the skylight - this will prevent any backed up water from leaking into the building

cross section

0437

Door cores

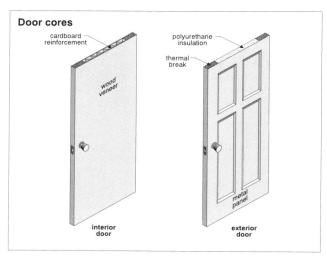

cardboard reinforcement

polyurethane insulation

thermal break

wood veneer

metal panel

interior door

exterior door

0438

Door operation

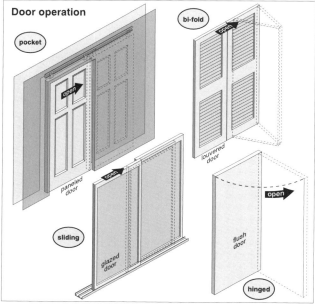

pocket

bi-fold

open

open

paneled door

louvered door

open

sliding

glazed door

flush door

hinged

0439

French doors

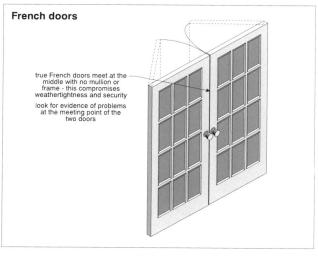

true French doors meet at the middle with no mullion or frame - this compromises weathertightness and security

look for evidence of problems at the meeting point of the two doors

0440

Insulated metal doors shouldn't have a storm door

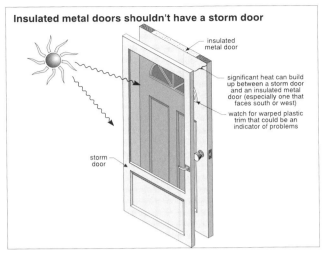

insulated metal door

significant heat can build up between a storm door and an insulated metal door (especially one that faces south or west)

watch for warped plastic trim that could be an indicator of problems

storm door

0 4 4 1

Doorsill support

look for a 1-1/2" to 6" step up into the house

this reduces the chance for water or snow to enter the house under the bottom of the door

apply weight here to check sill support

doorsill

6"

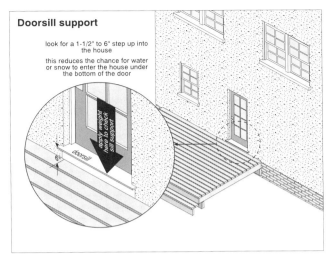

0 4 4 2

Staining below exterior doors

leakage through exterior doors is common

check for staining on the floor by the door and below the floor, if possible

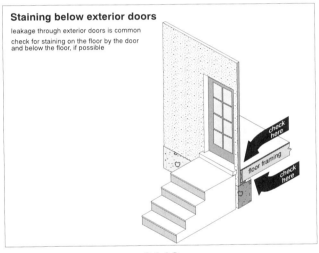

check here

floor framing

check here

0 4 4 3

Man door (attached garage)

garage

self closer

must be exterior-type door (fire rated in some areas)

house interior (not bedroom)

6" step up into house

door should be tight fitting and weatherstripped

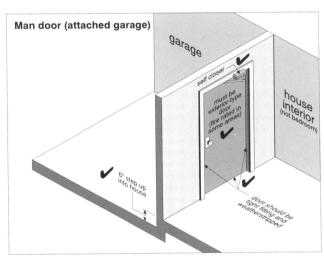

0 4 4 4

Causes of wet basement problems

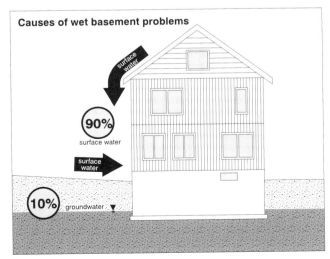

surface water

90% surface water

surface water

10% groundwater

0 4 4 5

Roof and surface water control

roof water control
gutters and downspouts carry the roof water to a safe discharge point

surface water control
surrounding ground should be graded down away from the house

48
1
minimum slope (non-permeable surface)
driveway

12
1
minimum slope (permeable surface)

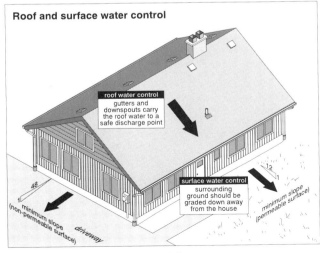

0 4 4 6

Basement windows and stairs

make sure that the grading is appropriate around basement walkouts and window wells

there should be provision for drainage

In some cases, covers may be necessary to keep the rain out

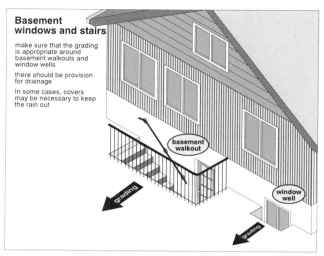

basement walkout

grading

window well

grading

0447

Swales

swales are shallow ditches that collect surface water several feet away from the building and divert it around one or both sides of the home

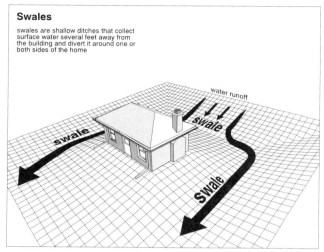

water runoff

swale

swale

swale

0448

Concrete flaws

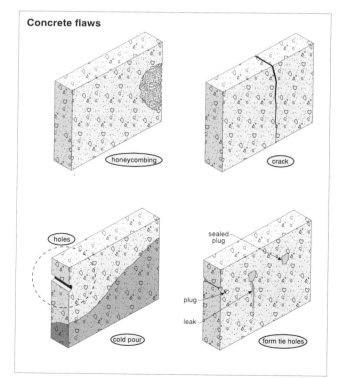

honeycombing

crack

holes

cold pour

sealed plug

plug

leak

form tie holes

0449

Where cracks appear

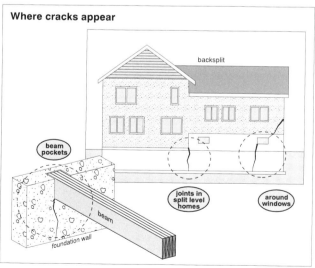

backsplit

beam pockets

beam

foundation wall

joints in split level homes

around windows

0450

Concrete block walls

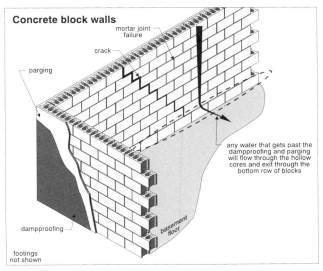

mortar joint failure

crack

parging

dampproofing

any water that gets past the dampproofing and parging will flow through the hollow cores and exit through the bottom row of blocks

basement floor

footings not shown

0451

High water table

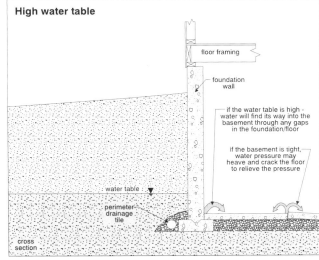

floor framing

foundation wall

if the water table is high - water will find its way into the basement through any gaps in the foundation/floor

if the basement is tight, water pressure may heave and crack the floor to relieve the pressure

water table

perimeter drainage tile

cross section

0452

Depth of foundation

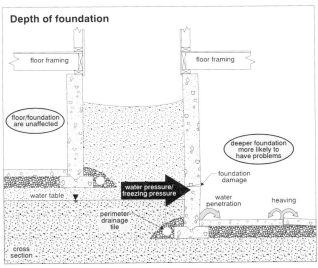

floor framing

floor framing

floor/foundation are unaffected

deeper foundation more likely to have problems

water pressure/ freezing pressure

foundation damage

water penetration

heaving

water table

perimeter drainage tile

cross section

0453

Wet basement clues - part 1

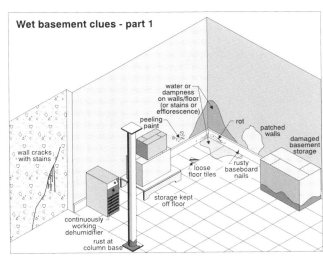

water or dampness on walls/floor (or stains or efflorescence)

peeling paint

rot

patched walls

damaged basement storage

wall cracks with stains

loose floor tiles

rusty baseboard nails

continuously working dehumidifier

storage kept off floor

rust at column base

0454

Wet basement clues - part 2

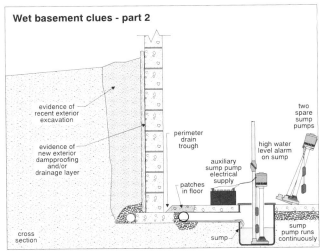

evidence of recent exterior excavation

evidence of new exterior dampproofing and/or drainage layer

perimeter drain trough

two spare sump pumps

high water level alarm on sump

auxiliary sump pump electrical supply

patches in floor

sump

sump pump runs continuously

cross section

0455

Why is there a sump pump in the basement?

there are reasons other than wet basement problems for having a sump pump in the basement

- the storm sewer may be higher than the perimeter drainage tile that discharges into the sump (not unusual with "infill" houses)
- in some areas local regulations may require a sump pump whether or not there is a problem

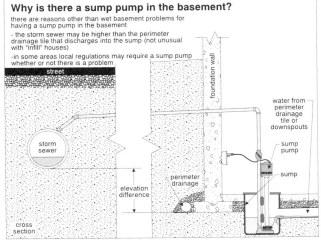

street

foundation wall

storm sewer

water from perimeter drainage tile or downspouts

sump pump

sump

elevation difference

perimeter drainage

cross section

0456

Control surface water

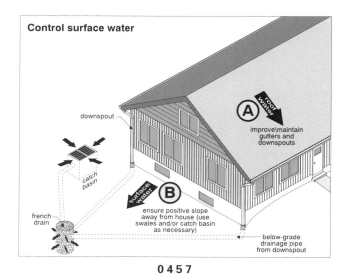

downspout

catch basin

french drain

A roof water

improve\maintain gutters and downspouts

B surface water

ensure positive slope away from house (use swales and/or catch basin as necessary)

below-grade drainage pipe from downspout

0457

Patching cracks

epoxy injection

injection tubes cemented into place - epoxy or polyurethane is then injected through the tubes

inside of foundation

grade level

drainage layer

outside of foundation

grade level

exterior patching

outside of foundation

grade level

drainage layer extending down to perimeter drainage tile

bituminous or rubberized asphalt dampproofing

0458

Interior drainage system

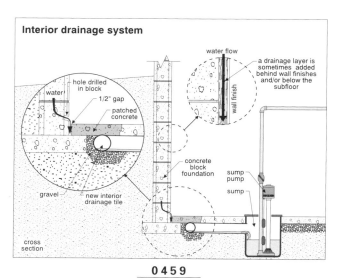

water flow

a drainage layer is sometimes added behind wall finishes and/or below the subfloor

hole drilled in block

water

1/2" gap

patched concrete

wall finish

concrete block foundation

sump pump

sump

gravel

new interior drainage tile

cross section

0459

Excavation, dampproofing and drainage tile

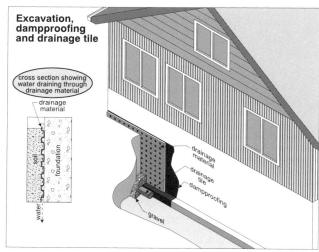

cross section showing water draining through drainage material

drainage material

soil

foundation

water

drainage material

drainage tile

dampproofing

gravel

0460

With over 500 illustrations, *Structure, Roofing and the Exterior* shows you exactly what the finished job should look like. This book is an excellent tool for the new home buyer or inspector, showing you which pitfalls to avoid and what to look for when searching for potential problems in the structure, roof or exterior of your or your client's home.

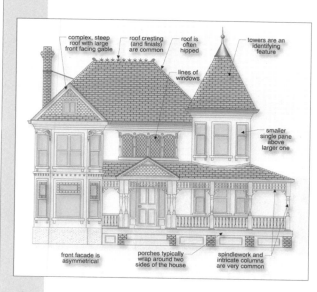

With almost 500 illustrations, *Heating and Air Conditioning* shows you exactly what the finished job should look like. This book is an excellent tool for the new home or inspector, showing you which pitfalls to avoid and what to look for when searching for potential problems in the heating and air conditioning systems of your or your client's

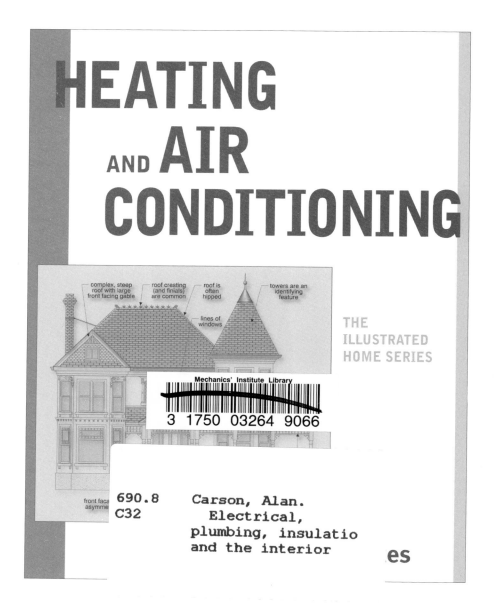

HEATING AND AIR CONDITIONING

THE ILLUSTRATED HOME SERIES

complex, steep roof with large front facing gable

roof cresting (and finials) are common

roof is often hipped

towers are an identifying feature

lines of windows

front faca asymme

690.8
C32

Carson, Alan.
   Electrical,
plumbing, insulatio
and the interior

es

MECHANICS' INSTITUTE